ADVANCED
HIGH SPEED
DEVICES

SELECTED TOPICS IN ELECTRONICS AND SYSTEMS

Editor-in-Chief: **M. S. Shur**

Selected Topics in Electronics and Systems – Vol. 51

ADVANCED HIGH SPEED DEVICES

Editors

Michael S. Shur
Rensselaer Polytechnic Institute, USA

Paul Maki
US Office of Naval Research, USA

 **World Scientific**

NEW JERSEY · LONDON · SINGAPORE · BEIJING · SHANGHAI · HONG KONG · TAIPEI · CHENNAI

Published by

World Scientific Publishing Co. Pte. Ltd.

5 Toh Tuck Link, Singapore 596224

USA office: 27 Warren Street, Suite 401-402, Hackensack, NJ 07601

UK office: 57 Shelton Street, Covent Garden, London WC2H 9HE

British Library Cataloguing-in-Publication Data
A catalogue record for this book is available from the British Library.

Selected Topics in Electronics and Systems — Vol. 51
ADVANCED HIGH SPEED DEVICES

Editor: Tjan Kwang Wei

ISBN-13 978-981-4287-86-9
ISBN-10 981-4287-86-5

Printed in Singapore by Mainland Press Pte Ltd

PREFACE

This volume contains the Proceedings of the 2008 biennial Lester Eastman Conference (LEC), which was held on the Cornell University of Delaware campus on August 5-7, 2008. Originally, the conference was known as the IEEE/Cornell University Conference on High Performance Devices. It was named to honor Prof. Lester Eastman, a renowned device pioneer and technology leader.

Professor Lester Eastman

The book covers five areas of advanced device technology: terahertz and high speed electronics, ultraviolet emitters and detectors, advanced III-V field effect transistors, III-N materials and devices, and SiC devices.

Very appropriately, the first paper in the issue is co-authored by Professor Lester F. Eastman. The paper presents experimental results on GaN based ultra-short planar negative differential conductivity diodes for THz power generation.

Reaching higher frequencies and higher powers using III-N materials system has become a hot topic. Monte Carlo simulations predict that III-N field effect transistors with nanometer gates should be able to penetrate the THz range of frequencies, and recent experimental results are encouraging. However, the extension of the effective gate length beyond the metallurgical gate, parasitic contact resistances and resistances of the gate-to-source and gate-to-drain regions adversely affect the high frequency performance of these devices. These issues are addressed in the paper by Simin et al who propose a novel five terminal design for THz GaN-based transistors with 10 nm gates and validate the design with ADS simulations.

A different approach in controlling short channel effects impeding high frequency operation is to use nanowire field effect transistors. Wang et al present performance comparison of scaled III-V and Si ballistic nanowire MOSFETs analysis and simulations of Si and III-V Gate-All-Around nanowire MOSFETS assuming ballistic or quasi-ballistic transport.

The next paper by Diduck et al proposes ballistic deflection transistor that uses the change in the current pathway. The authors consider possible operation of such devices at room temperature.

Otsuji et al present an excellent review of using plasma waves – waves of electron density – in dual-grating-gate HEMTs for emission and intensity modulation of terahertz electromagnetic radiation. This innovative approach has already led to the observation of THz emission from short channel HEMTs at room temperature.

Ken O et al discuss the feasibility of CMOS circuits operating at frequencies in the upper millimeter wave and low sub-millimeter frequency regions. They refer to the demonstrated 140-GHz fundamental mode VCO in 90-nm CMOS, a 410-GHz push-push VCO in 45-nm CMOS, and a 180-GHz detector circuit in 130-nm CMOS have been demonstrated and conclude that, with the continued scaling of MOS transistors, 1-THz CMOS circuits will be possible.

Several papers in the Proceedings are devoted to with ultraviolet light emitting diodes (UV LEDs) and detectors. Sampath et al discuss the effects of nanometer scale compositional inhomogeneities in the active regions of UV LEDs with high Al mole fraction in AlGaN active regions. They report on prototype flip chip double heterostructure UV LEDs operating at 292 nm.

Chivukula et al demonstrate a strong effect of pulsed sub-band ultraviolet illumination on surface acoustic wave propagation in GaN-on-sapphire.

The paper by Alexey Vert et al on solar-blind single-photon 4H-SiC avalanche photodiodes reports on the record performance. The paper shows that SiC UV photodetectors can successfully compete and even outperform III-N based UV photodiodes.

The next section of the book deals with advanced III-V Field Effect Transistors. Ayubi-Moak et al present the results of the Monte Carlo simulations of $In_{0.75}Ga_{0.25}As$ MOSFETs at 0.5 V supply voltage for high-performance CMOS operation.

Karimy et al describe the first 70 nm 6-inch GaAs PHEMT MMIC process. This millimeter wave technology demonstrated excellent performance from Ka-band through W-bands. The device DC and RF characteristics have excellent uniformity across the wafer.

Dong Xu et al report on high-performance 50-nm metamorphic high electron-mobility transistors with high breakdown voltages. This has been achieved by the optimization of the epitaxial layer design (including a high indium composite channel and the double-sided doping), by the selection of the proper gate recess scheme, and by using an asymmetric gate recess. Their results demonstrate that these devices are excellent candidates for ultra-high-frequency power applications.

Papers on III-N materials and devices are included into the next section, which starts from the paper by Chen et al MBE growth and characterization of Mg-doped III-nitrides on sapphire.

Ke Tang et al. discuss the performance of MOSFETs on reactive-ion-etched GaN surfaces. They report on field effect mobilities reaching 170 cm^2/V-s and subthreshold slope of 3.8 V/decade for as grown GaN MOSFETs.

Shi et al present new results for high current density/high voltage AlGaN/GaN HFETs on sapphire. For a gate-drain spacing of 10μm, they achieved a specific on-resistance of 1.35mΩ-cm^2 and off-state breakdown voltage of 770V.

M. Alomari et al report on InAlN/GaN MOS-HEMT with thermally grown oxide. The gate leakage current was reduced by two orders of magnitude, and pulse measurements showed lag effects similar to those for devices without oxidation, indicating a high quality native oxide. The MOS-HEMT yielded a power density of 5 W/mm at 30 V drain voltage at 10 GHz, power added efficiency of 42% and F_t and F_{max} of 42 and 61 GHz respectively.

Tetsuzo Ueda et al. review their state-of-the-art GaN-based device technologies for power switching at low frequencies and for future millimeter-wave communication systems. They established crack-free epitaxial growth of GaN on Si and proposed a novel device structure called Gate Injection Transistor (GIT) that achieved normally-off operation with high drain current. Short-gate MIS-HFETs using in-situ SiN as gate insulators achieved f_{max} up to 203GHz and enabled compact 3-stage K-band amplifier MMIC with the small-signal gain is as high as 22dB at 26GHz.

Zimmermann et al reported on 4-nm AlN barrier all binary MISHFETs with SiN_x gate dielectric. They achieved very low sheet resistances, ~ 150 Ohm/sq, a high carrier mobility and concentration (~ 1200 cm^2/Vs and ~ 3.5×10^{13} cm^{-2} at room temperature) and output current densities of 1.7 A/mm and 2.1 A/mm with the intrinsic transconductances of 455 mS/mm and 785 mS/mm for 2 µm and 250 nm gate-length devices, respectively

The last section of the book deals with SiC Devices. Naik, Tang, and Chow report on effect of gate oxide processes on 4H-SiC MOSFETs. Naik, Wang, and Chow discuss characterization and modeling of integrated diode in 1.2kV 4H-SiC Vertical Power MOSFET. O'Brien and Koebke consider packaging and wide-pulse switching of 4 mm x 4mm silicon carbide GTOs. And, finally, Urciuoli and Veliadis report on bi-directional scalable solid-state circuit breakers for hybrid-electric vehicles

The conference was under the technical sponsorship of the Institute of Electrical and Electronic Engineering (IEEE). We are grateful to the National Science Foundation, Air Force Office of Scientific Research (AFOSR), the Office of Naval Research (ONR), and the University of Delaware for their support of the Lester Eastman Conference, 2008.

EDITORS
Michael S. Shur (shurm@rpi.edu)
Paul Maki (Paul_Maki@onr.navy.mil)

CONTENTS

International Journal of High Speed Electronics and Systems
Vol. 19, No. 1 (2009) 1–6
© World Scientific Publishing Company

SIMULATION AND EXPERIMENTAL RESULTS ON GaN BASED ULTRA-SHORT PLANAR NEGATIVE DIFFERENTIAL CONDUCTIVITY DIODES FOR THz POWER GENERATION

BARBAROS ASLAN

*School of Electrical and Computer Engineering, Cornell University, 424 Phillips Hall,
Ithaca, NY 14853, USA
ba58@cornell.edu*

LESTER F. EASTMAN[1], QUENTIN DIDUCK[2]

*[1] School of Electrical and Computer Engineering, Cornell University, Ithaca, NY 14853, USA
[2] School of Electrical and Computer Engineering, University of Rochester, Rochester, NY 14627, USA*

A GaN based negative differential conductivity diode utilizing transient ballistic transport effects is proposed and large-signal circuit simulations along with preliminary experimental results are presented. The diode is an n^+-n-n^+ structure and transport is described by an empirical velocity-field relation which is derived directly from femtosecond pulse-probe measurements available in literature and incorporated into the simulations through curve fitting. Efficient THz generation is predicted as a result of ~2.8 peak-to-valley ratio. Pulsed current-voltage characteristics were measured and N-type dependence was observed.

Keywords: Terahertz; Ballistic Transport; Negative Differential Conductivity; Negative Differential Resistance; Planar Diode; GaN Diode

1. Introduction

A number of novel approaches have been suggested to fill the THz gap, a frequency range that is difficult to cover neither with optical nor with electronic devices. Quantum cascade lasers [1] show promising results, producing mW of CW power; however room temperature operation still remains a major challenge. Terahertz emission from ultra-short gate FETs at room temperature has recently been detected in which plasma wave generation due to Dyakonov-Shur instability is responsible [2] and more theoretical and experimental work is on the way to optimize this phenomena. In this article however, we propose a different device approach. It relies on the strong transient ballistic transport properties of electrons drifting in an ultra-short GaN channel under the influence of strong bias fields. This leads to velocity reduction with increasing field and creates a strong negative differential conductivity (NDC) which is fast enought to generate THz radiation. We call this device the BEAN diode (Ballistic Electron Acceleration Negative differential condutivity). Electron transport for this device is described by an empirical velocity-field characteristic derived from the femtosecond pulse-probe experiments of Wraback et al. [3, 4] and is shown in Fig. 1 (solid triangles)

Using light pulses with photon energy just above the GaN bandgap, he was able to measure the electron accumulation layer drift velocity vs. time in a micron long GaN sample, biased at series of electric field strenghts. Most of these transient effects occured in the first ~250 nm drift region. Therefore, by selecting a 250 nm long channel, the average transit velocity can be determined at this and other bias field strengths. This yields the average drift velocity vs. electric field for such a channel as shown in Fig. 1. Intervalley transition time in GaN has been measured by Wu et al [5] and is reported to be comparable to their 0.17 ps pump time. Furthermore, close inspection of Wraback et al's velocity-time measurements at high fields indicates that the drift velocity drops rapidly within ~0.3 ps after the peak value of ~7×10^7 cm/s has been reached. This value is made longer than its real value due to the 0.07 ps pulse and 0.1 ps probe times in his experiments. Therefore, his data roughly substantiates Wu et al.'s ~0.17 ps as the intervalley transition time. This time constant, together with the the acceleration time at the end of which the electron reaches at or above ~1 eV kinetic energy (enabling the electrons to transfer), determine the 3dB cut off frequency. This acceleration time can be estimated from the acceleration law $\hbar \dot{k} = qE$ as ~50 fs using the value of k as the boundary condition at which $E(k) \approx 1\,eV$. Therefore, the 3 dB cut off condition f=1/τ, requires τ = 56 + 170 = 226 fs, which corresponds to a frequency of ~4.5 THz. There may be, in addition to electron transfer in Wraback et al's data, also a modest amount of phonon build up and even electron negative effective mass effect responsible in the velocity reduction observed. Thus, through the transient ballistic transport effects, it is possible to create NDC with a very fast time response.

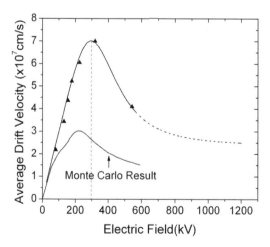

Fig. 1. Drift velocity versus electric field constructed from experimental measurements. Solid curve is a fit to the discrete experimental data. Also shown in continuous line is a typical Monte Carlo result to contrast the transient ballistic effects involved in transport over short distances.

2. Simulation Results

The diode is modeled as one-dimensional n^+-n-n^+ structure with a doping profile of 1×10^{20}, 1×10^{18} and 1×10^{20} cm^{-3}, respectively. Previously, other simulation results have been reported for long n+-n-n+ channels [6,7].

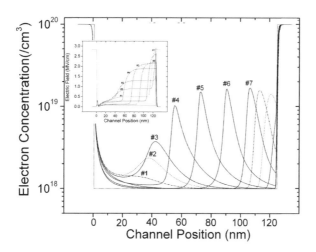

Fig. 2. Electron concentration vs. position at different instance of time (numbered labels) showing accumulation layers in transit towards the anode. Inset shows the accompanying spatial distribution of the electric field.

The n^+ layers are important for good ohmic contacts and also serve to provide electron concentration gradient for the nucleation of accumulation layers. The channel doping is chosen to meet the doping-length product criterion, NL, to allow space charge instabilities and also is crutial for the generation of a useful a.c. power level. Poisson and current continuity equations are solved within the framework of drift-diffusion approximation using physics-based device simulator "ATLAS" (Silvaco Inc.). Diffusion coefficient is calculated for each electric field value using the Einstein's relationship.

The applicability of the derived velocity-field characteristics at 1×10^{18} cm^{-3} doping level, (which was constructed from measurements performed in a 1×10^{15} cm^{-3} doped sample) is justified by the fact that the hot electrons, as is the case in these diodes under bias, are only minimally affected by the impurity scattering. This has been verified by Foutz et al. in GaN through Monte Carlo simulations [8]. Moreover, from a fabrication perspective, 1×10^{18} cm^{-3} is appropriate since typical molecular beam epitaxy (MBE) doping profiles are significantly more reproducible above ~5×10^{17} cm^{-3}.

To ensure accuracy, time steps have been chosen to be much less than the dielectric relaxation time, $\Delta t \ll \tau_\varepsilon$, and likewise, spatial mesh size much less than $v\tau_\varepsilon$ where v is the average velocity of the electrons [9]. The diode is assumed to be in parallel with a parallel RLC circuit where C represents the geometric capacitance of the diode (4.2 fF),

R is the load resistor (220 Ω) and L is the inductor to create resonance (0.61 pH). The geometric capacitance is dominated by the fringing fields penetrating GaN due to ~10:1 ratio of its dielectric constant to air. The device is designed 50 µm wide with a channel thickness of 20 nm. When biased above the threshold voltage V_{TH}, accumulation layers nucleate and are periodically drawn from the n$^+$ layer at the cathode and propagate towards the anode where they are collected as shown in Fig. 2. This leads to current-voltage oscillations at the terminals of the diode at the transit time frequency.

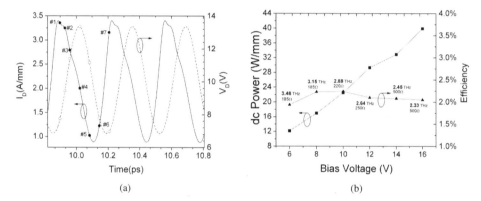

Fig. 3. (a) Current-Voltage waveforms as a function of time. (b) Efficiency and d.c. Power dissipation vs. bias voltage for a 125 nm device. Frequency and the optimum load resistors are also labeled.

This particular simulation was performed for 10 V bias. The output wave forms were Fourier analyzed and following performance parameters were calculated: 2.9 THz. fundamental frequency, 0.5 W/mm a.c. output power (~25 mW) at 2.22% conversion efficiency, 6.68 V peak-to-peak voltage swing amplitude, 1.96 A/mm peak-to-peak current swing amplitude. The current-voltage oscillations at the terminals are shown in Fig. 3a. In Fig. 3b, bias dependency of the conversion efficiency is shown. For each bias point, load resistor and the parallel inductor has been tuned in order to maximize the efficiency.

3. Experimental results and discussion

In order to verify simulation results, a device with contact spacing of ~125 nm was produced. This device was fabricated on GaN bonded to a polycrystalline diamond substrate for thermal management. The GaN growth contained a 40 nm thick n-type epilayer doped at 1×10^{18} cm^{-3} and a 190 nm thick n$^+$ region doped at 1×10^{20} cm^{-3} in order to reduce ohmic contact resistance. Several devices were fabricated and exhibited negative differential conductivity (NDC) characteristics. The fabrication required removal of the n+ region to properly form the channel of the device. During fabrication a thin n+ region, approximately 10 angstroms thick, remained after the etch that caused an undesirable shunt conductance.

Shown in Fig. 4 is the measured result of the device under 200 ns pulsed I-V measurements, also shown is an estimated correction for the shunt conductance. A large peak to valley ratio is expressed in this device and is on the order of the expected result based upon Wraback's data [3, 4] shown in Fig. 1. The approximate field strength of the peak velocity is at approximately 300 kV/cm, based upon the etch angle of the channel and assuming mild ohmic conductive losses. Due to the shunt conductance the devices failed shortly after measurement, but several devices displayed near identical characteristics though with slightly smaller peak to valley ratio's than is shown in Fig. 4. While these results are initial, we expect that reliable devices will be produced in the near future.

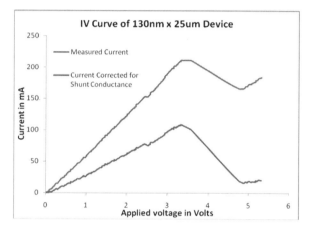

Fig. 4. The top plot indicated the measured results including an undesirable shunt conductance due to a thin n+ region that was not completely etched. The bottom plot is an estimated correction based upon a 10 angstrom thick n+ layer.

4. Conclusion

Large signal simulation results of GaN NDC diodes based on an empirical velocity-field relation is presented and efficiencies up to 2.3 % is shown to be possible utilizing the transient transport effects in ultra-short distances. To implement these devices, a fabrication technique has been developed and preliminary results of pulsed I-V measurement of a ~125 nm device have indicated N-type dependence.

References

[1] L. Mahler, R. Kohler, A. Tredicucci, F. Beltram, H. E. Beere, E. H. Linfield, D. A. Ritchie and A. G. Davies, "Single-mode operation of terahertz quantum cascade lasers with distributed feedback resonators," *Appl.Phys.Lett.*, vol. 84, no. 26, pp. 5446-5448, 2004.
[2] N. Dyakonova, A. El Fatimy, J. Lusakowski, W. Knap, M. I. Dyakonov, M. -. Poisson, E. Morvan, S. Bollaert, A. Shchepetov, Y. Roelens, C. Gaquiere, D. Theron and A. Cappy, "Room-temperature terahertz emission from nanometer field-effect transistors," *Appl.Phys.Lett.*, vol. 88, no. 14, pp. 141906, 2006.

[3] M. Wraback, H. Shen, J. C. Carrano, C. J. Collins, J. C. Campbell, R. D. Dupuis, M. J. Schurman and I. T. Ferguson, "Time-resolved electroabsorption measurement of the transient electron velocity overshoot in GaN," *Appl. Phys. Lett.,* vol. 79, no. 9, pp. 1303-5, 08/27. 2001.

[4] M. Wraback, H. Shen, S. Rudin, E. Bellotti, M. Goano, J. C. Carrano, C. J. Collins, J. C. Campbell and R. D. Dupuis, "Direction-dependent band nonparabolicity effects on high-field transient electron transport in GaN," *Appl. Phys. Lett.,* vol. 82, no. 21, pp. 3674-3676, 2003.

[5] S. Wu, P. Geiser, J. Jun, J. Karpinski, D. Wang and R. Sobolewski, "Time-resolved intervalley transitions in GaN single crystals," *J. Appl. Phys.,* vol. 101, no. 4, pp. 043701, 2007.

[6] R. F. MacPherson, G. M. Dunn and N. J. Pilgrim, "Simulation of gallium nitride Gunn diodes at various doping levels and temperatures for frequencies up to 300 GHz by Monte Carlo simulation, and incorporating the effects of thermal heating," *Semiconductor Science and Technology,* vol. 23, no. 5, pp. 055005, 2008.

[7] P. Shiktorov, E. Starikov, V. Gruzinskis, M. Zarcone, D. Persano Adorno, G. Ferrante, L. Reggiani, L. Varani and J. C. Vaissiere, "Monte Carlo analysis of the efficiency of Tera-Hertz harmonic generation in semiconductor nitrides," *Physica Status Solidi (A) Applied Research,* vol. 190, no. 1, pp. 271-279, 2002.

[8] B. E. Foutz. "Electron transport and device modeling in Group-III nitrides," Ph.D dissertation, Cornell University, Ithaca, NY.

[9] P.J. Bulman, G.S. Hobson and B.C. Taylor, *Transferred electron devices,* New York: Academic Press, 1970.

International Journal of High Speed Electronics and Systems
Vol. 19, No. 1 (2009) 7–14
© World Scientific Publishing Company

5-TERMINAL THz GaN BASED TRANSISTOR WITH FIELD- AND SPACE-CHARGE CONTROL ELECTRODES

GRIGORY SIMIN[1], MICHAEL S. SHUR[2], REMIS GASKA[3]

[1]*Department of Electrical Engineering, University of South Carolina, 301 S. Main street.,*
Columbia, SC, 29208, USA;
[2]*Center for Integrated Electronics, Rensselaer Polytechnic Instutite, 110 8th Street*
Troy, New York 12180, USA, shurm@rpi.edu
[3]*Sensor Electronic Technology Inc., 1195 Atlas Road, Columbia, SC 29209, USA*

We present a novel approach to achieve terahertz-range cutoff frequencies and maximum frequencies of operation of GaN based heterostructure field-effect transistors (HFETs) at relatively high drain voltages. Strong short-channel effects limit the frequency of operation and output power in conventional geometry GaN HFETs. In this work, we propose a novel device with two additional independently biased electrodes controlling the electric field and space-charge close to the gate edges. As a result, the effective gate length extension due to short channel effects is diminished and electron velocity in the device channel is increased. Our simulations show that the proposed five-terminal HFET allows achieving f_T=1.28 THz and f_{max}= 0.815 THz at the drain voltages as high as 12 V. Hence, this device opens up a new approach to designing THz transistor sources.

Keywords: HFETs; THz sources; Gallium Nitride; field-control electrodes

1. Introduction

Electronic devices operating in the THz frequency range will enable applications in radio astronomy, Earth remote sensing, radars and vehicle radars, scientific investigations, non-destructive testing of materials and electronic devices, chemical analysis, explosive detection, moisture content, coating thickness control, imaging, and wireless covert communications. In spite of great demand for efficient THz sources, compact and efficient transistor THz sources are not yet available. Achieving electronic device operation in the THz range is a complex multifaceted problem involving the control of electron velocity, electric field and potential profiles, access resistances, parasitic parameters, and electromagnetic coupling. Figure 1 clearly shows that electronic sources run out of steam above 0.1 THz, whereas photonic THz sources are more effective at frequencies above 10 THz. In the THz range, current electronic sources, such as photomixers and frequency multipliers can only deliver the RF powers in micro-watt range. THz lasers can emit high powers, up to 1 W, however, these devices are bulky, require high pumping powers and cannot be fabricated using integrated technology, which is a key feature of the modern functional system and building block design concept.

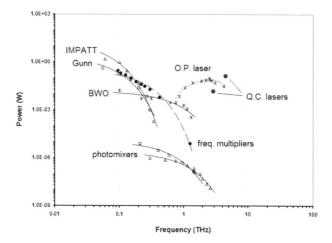

Figure 1. Illustration of THz gap problem (from [1] IEEE©2007). Filled circles correspond to cryogenic sources

One of the most important criteria for an efficient THz emission is the peak electron velocity.

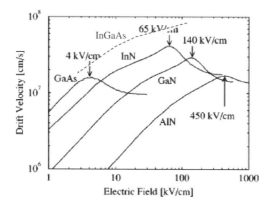

Figure 2. Drift electron velocity in various III-V materials (after [2, 3])

Figure 2 compares electron velocity- field dependencies in various III-V materials. InN (4.5×10^7 cm/s) and InGaAs (close to 10^8 cm/s) have the highest peak velocities. GaN is the next to these materials with the peak velocity of 3×10^7 cm/s. In very short-gate devices, the average velocities under the gate might be considerably higher than the steady state values, due to so-called overshoot effects. [2]. Both InN and GaN have peak overshoot velocities are expected to be close to 10^8 cm/s. However, in InN material, the electron reach peak velocities at lower electric fields thus allowing for highest average velocities at longer gate length, around 0.15 μm as compared to 0.05 μm gate length required for GaN based devices.

Recently the cutoff frequencies in THz or sub-THz range have been demonstrated by a number of groups. The f_{max} above 1 THz has been achieved with InGaAs/InP HEMT [4], 300 GHz InP HEMT MMIC with 100 µW output power has been demonstrated by HRL. [5] Outstanding results have been achieved using advances in Si-technology. NMOS and PMOS devices fabricated using 45nm gate process with f_T of 485 GHz and 345 GHz respectively have been demonstrated by IBM. [6] The achievements in THz sources development correspond to the cutting edge III-V and Si technology; yet no THz operation and high-powers have been achieved simultaneously. The key obstacles to solving the problems remain relatively low current density in Si, GaAs and InP based devices, rapid degradation of cut-off frequencies with increasing drain bias (due to short-channel effects), and low operating voltages in devices with submicron inter-electrode spacing.

2. GaN Heterostructure Field-Effect Transistors

GaN based Heterostructure field-effect transistors (HFETs, a.k.a. HEMTs) exhibit record high electron densities and high peak electron velocity and mobilities in the 2D channel, which promise simultaneous achievement of cut-off frequencies, f_T, in the THz range and high output powers using nanoscale (~ 30 nm) gate technology. This makes GaN HFETs excellent candidates for high-power solid-state THz sources. The key problems precluding THz operation of GaN HFETs are related to effective gate length increase at high drain bias and access resistances causing significant degradation in f_T and f_{max} frequencies.

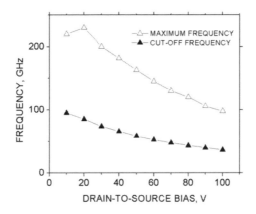

Figure 3. Cut-off frequencies of GaN based HFET vs. drain bias (from [7])

Figure 3 shows the dependence of f_T and f_{max} for GaN HFET with 150 nm long gate reported in [7]. The simulation results of [7] lead to important observation illustrating the difficulties in achieving THz operation with GaN HFETs. The effective gate length significantly exceeds the physical gate length. The difference is due to the 2DEG space charge region expansion into the gate-to-drain spacing with increasing drain bias. For the device with the actual gate length of 0.15 µm, the effective gate length reaches 0.25 µm at

14 V drain bias and around 0.5 μm at 32 V drain bias. It has been suggested that an additional field-controlling electrode (FCE) located in the near vicinity of the drain-side gate edge can efficiently control the extend of the space charge and thus the cut-off frequencies at high drain bias. Figure 4 shows the experimental data on the f_T – drain bias dependence for GaN HFETs with the drain FCE.

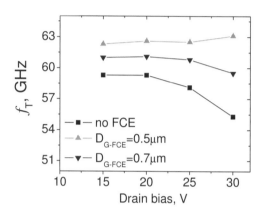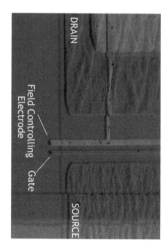

Figure 4. The F_T – drain bias dependencies for GaN HFET with 0.3 μm long gate and different gate-to FCE spacing (after [8])

As seen the addition of the FCE on the drain side of the gate allows for a complete suppression of the effective gate length increase and correspondingly leads to the nearly bias independent f_T. Another important set of limitations on the microwave performance of GaN HFETs arises from the access resistances. It is well known that, in GaN technology the achievable contact resistance values are significantly (around an order of magnitude) higher than those in GaAs technology. Contact annealing in GaN devices also requires much higher temperatures, leading to rough contact edges and calling for larger gate – to ohmic spacing to avoid premature breakdown and inter-electrode shortening. The source access resistance R_S comprising contact resistance R_C and source-gate opening resistance R_{SG} significantly reduces the external transconductance of sub-μm gate devices and leads to lower drain saturation currents. In addition, the total source and drain access resistances increase the knee voltage, thus requiring higher drain voltage to operate the device and to achieve high RF powers. Another problem associated with the access region is the 2DEG depletion due to the surface potential modulation. As a result, the carrier concentration in channel outside the gate becomes lower than that under the gate at high positive input signals. This leads to lower power gain and to an increase in the effective gate length at large input signals.

A new approach to significantly decreasing the contact resistance and allowing for a very tight source-gate-drain spacing was proposed in [9, 10]. This approach uses capacitively-coupled contacts (C^3) to fabricate microwave HFETs with low contact resistance at microwave frequencies and with independent control of induced carrier concentration in the source-gate and gate drain openings. [9]

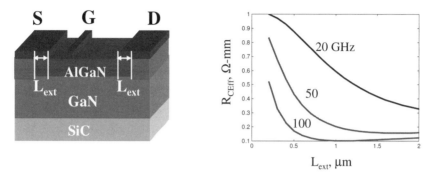

Figure 5. Schematic structure and effective contact resistance of the RF-enhanced contact to GaN HFET [9]

3. Proposed novel five-terminal GaN based THz HFET

In this work, we propose a novel device with two additional independently biased electrodes controlling the electric field and space-charge in the close vicinity of the gate edges, both at the source and drain sides . Source and drain field- and space-charge field controlling electrodes (FCEs) capacitively coupled with the source and drain ungated regions fundamentally change the HFET performance at THz frequencies. Additional bias voltages applied to the source and FCEs increase the electron concentration and velocity at the source edge of the gated channel and control the space-charge spread at the drain edge. As a result, the electron velocity under the gate electrode is increased and the space-charge penetration into the gate-drain region is minimized.

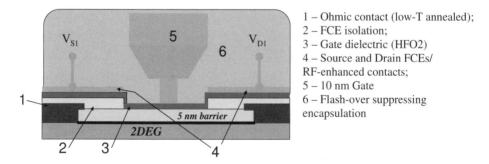

1 – Ohmic contact (low-T annealed);
2 – FCE isolation;
3 – Gate dielectric (HFO2)
4 – Source and Drain FCEs/ RF-enhanced contacts;
5 – 10 nm Gate
6 – Flash-over suppressing encapsulation

Figure 6. Schematic structure of 5-terminal THZ HFET with field-controlling electrodes (FCEs)

Due to very short gate lengths in THz HFETs, even small values of the source access resistance have significant effect on the device peak current and therefore, on the transconductance and cut-off frequency. The effect of source resistance on the

MOSHFET with the gate length $L_G = 0.03$ μm is illustrated in Fig. 7 (a). The threshold voltage $V_T = -3$ V was used in these simulations.

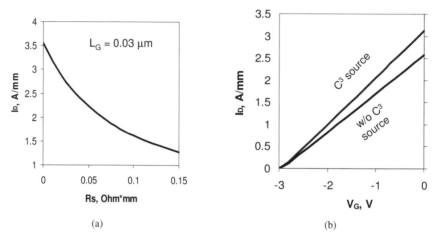

(a)   (b)

Figure 7. (a): Effect of the source access resistance on the MOSHFET peak drain current; (b): Transfer characteristics of the MOSHFET with and without source C^3 electrode.

The effect of source – drain access resistance reduction due to capacitively-coupled source electrode is illustrated in transfer characteristics of Fig. 7 (b). The transfer characteristics were simulated using the MOSHFET with the gate length $L_G = 0.03$ μm and the source-gate spacing of 0.1 μm. The AlGaN barrier thickness was 10 nm. For the device without C^3 source electrode, the access resistance of the source-gate spacing was calculated using the sheet resistance $R_{SH} = 300$ Ω/sq. For the device with C^3 source electrode under positive bias inducing additional electrons into the 2 D channel, we used the data of [11] providing the maximum additional carrier concentration that can be induced in the AlGaN/GaN 2DEG channel. According to [11], the induced carriers can reduce R_{SH} by a factor of two. The value of R=150 Ω/sq was used to simulate the transfer curve for the MOSHFET with source C^3 electrode in Fig. 7 (b).

The ADS simulations for the proposed 5-terminal GAN HFET at 10 V drain bias are shown by solid lines in Figure 8. For the ADS simulations we used a MOSFET level 3 model to simulate the intrinsic HFET. The model input data were recalculated using parameters specific for the GaN HFET, such as 2DEG equilibrium sheet density, threshold voltage, barrier and dielectric thickness etc. The access regions were simulated using equivalent circuit of the RF-enhanced contacts extracted from MATLAB simulations (the data presented in Figure 5). Physical gate length was taken as 0.01 μm. The effective gate length for the device without FCE, was taken as drain bias dependent, increasing from the geometrical value at zero drain bias to 80 nm at 10 V drain bias according to the data from reference [7]. For the 5-terminal device with FCEs, the 0.02 μm gate FCE separation was assumed. Correspondingly, the effective gate length was kept at 30 nm independently of the drain bias as the gate depletion region extension is controlled by the gate – FCE spacing. The effective electron velocity was estimated using the technique and results of [7]. For a conventional geometry HFET, the electron velocity was $v_m = 1.5 \times 10^7$ *cm/s* whereas for the HFET with FCEs, $v_m = 2.7 \times 10^7$ *cm/s* (see Fig. 1.).

As seen from the simulation results of Figure 8, the novel approach allows achieving $f_T = 1.28$ THz and $f_{max} = 0.815$ THz at the drain voltages as high as 10 V. The proposed device, therefore, opens up a new approach to fabricating high-power THz transistor sources.

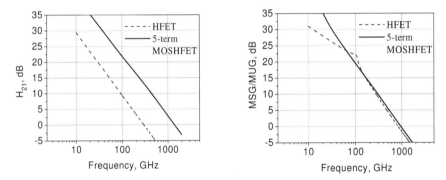

Figure 8. Current and power gain of 5-terminal HFET simulated with ADS.

Table 1 shows the summary of the input device parameters and performance of the proposed 5-terminal HFETs in comparison with the conventional geometry HFET.

Table 1

Parameter	5-Term. HFET	Regular HFET
Layout	$L_G = 10$ nm; W = 15 um;	
Effective gate length	$L_G = 30$ nm	$L_G = 80$ nm
Electron velocity	$2.7 \ 10^7$ cm/s	$1.5 \ 10^7$ cm/s
f_T	1.28 THz	300 GHz
f_{MAX}	815 GHz	700 GHz

Expected output powers of the proposed device can be estimated as follows. Consider the 5-terminal MOSHFET with the total width of 20 μm. The drain bias of 5 V can be applied across this device assuming that the surface flash-over breakdown effects are suppressed by the dielectric encapsulation. In this case, the highest electric field existing across the gate – FCE region, is 5V/0.02 μm is 2.5 MV/cm, not exceeding the breakdown filed for GaN. The peak drain current of such device, according to the data of Reference [11], is $I_M \approx 3.5$ A/mm $\times$ 20×10^{-3} mm = 70 mA. The output power at the drain bias $V_D = 5$ V and the knee voltage $V_{KN} = 2$ V is $P_m \approx I_M \times (V_D - V_{KN})/4 \approx 35$ mW. This power is much higher than that ever obtained using InGaAs-based or other THz transistors.

In conclusion, we considered the possibility of using high electron velocities in III-N materials to penetrate the THz range with higher power than InGaAs based devices. We analyzed the factors preventing the III-N devices from achieving THz cut-off frequencies. These include effective gate length extension and high access resistances. We proposed a novel 5-terminal device with field-controlling electrodes which has a promise to achieving the current cutoff frequency of 1.28 THz and f_{max} of 815 GHz at the drain bias as high as 10 V.

4. Acknowledgement

The work at RPI has been supported by the National Science Foundation under the auspices of I/UCRC "Connection One." The work at SET Inc. and USC was supported by the SBIR Phase I contract monitored by A. L. de Escobar, E. Wong and W. Kordela. The work at USC was supported by the National Science Foundation under Grant No. 0621931 monitored by Dr. P. Fulay.

References

[1] W.J. Stillman and M.S. Shur, Closing the Gap: Plasma Wave Electronic Terahertz Detectors, Journal of Nanoelectronics and Optoelectronics, Vol. 2, Number 3, pp. 209-221, December 2007

[2] B. E. Foutz, S. K. O'Leary, M. S. Shur, and L. F. Eastman, J. Appl. Phys. 85, 7727 (1999)

[3] v(E) for InAs (from Hess and Brennan (1984), See M.P. Mikhailova Handbook Series on Semiconductor Parameters, vol.1, M. Levinshtein, S. Rumyantsev and M. Shur, ed., World Scientific, London, 1996, pp. 147-168.

[4] R. Lai, X. B. Mei, W.R. Deal, W. Yoshida, Y. M. Kim, P.H. Liu, J. Lee, J. Uyeda, V. Radisic, M. Lange, T. Gaier, L. Samoska, A. Fung, Sub 50 nm InP HEMT Device with Fmax Greater than 1 THz, IEDM Technical Digest, p. 609 (2007)

[5] http://www.hrl.com/html/techs_mel.html

[6] Sungjae Lee, Basanth Jagannathan, Shreesh Narasimha, Anthony Chou, Noah Zamdmer, Jim Johnson, Richard Williams, Lawrence Wagner, Jonghae Kim, Jean-Olivier Plouchart, John Pekarik, Scott Springer and Greg Freeman, Record RF performance of 45-nm SOI CMOS Technology, IEDM Technical Digest, p. 225 (2007)

[7] V. O. Turin, M. S. Shur, and D. B. Veksler, Simulations of field-plated and recessed gate gallium nitride-based heterojunction field-effect transistors, International Journal of High Speed Electronics and Systems, vol. 17, No. 1 pp. 19-23 (2007)

[8] Pala, N. Yang J., Z. Koudymov, A. Hu, X. Deng, J. Gaska, R. Simin, G. Shur, M. S. Drain-to-Gate Field Engineering for Improved Frequency Response of GaN-based HEMTs, Device Research Conference, 2007 65th Annual, 18-20 June 2007, pp. 43-44

[9] G. Simin, Wide Bandgap Devices with Non-Ohmic Contacts, 210th Electrochemical Society Meeting 2006, Cancun, Mexico October 29-November 3, 2006

[10] G. Simin, Z-J. Yang, M. Shur, High-power III-Nitride Integrated Microwave Switch with capacitively-coupled contacts, Microwave Symposium, IEEE/MTT-S International, pp. 457-460 (2007)

[11] A. Koudymov, H. Fatima, G. Simin, J. Yang, M. Asif Khan, A. Tarakji, X. Hu, M. S. Shur, and R. Gaska Maximum Current in Nitride-Based Heterostructure Field Effect Transistors Appl. Phys. Lett, V. 80 pp. 3216-3218 (2002)

International Journal of High Speed Electronics and Systems
Vol. 19, No. 1 (2009) 15−22
© World Scientific Publishing Company

PERFORMANCE COMPARISON OF SCALED III-V AND Si BALLISTIC NANOWIRE MOSFETs

LINGQUAN (DENNIS) WANG

Electrical and Computer Engineering, University of California, San Diego
La Jolla, CA 92093,USA
liw001@ucsd.edu

BO YU

Electrical and Computer Engineering, University of California, San Diego
La Jolla, CA 92093,USA
boyu@ucsd.edu

PETER M. ASBECK

Electrical and Computer Engineering, University of California, San Diego
La Jolla, CA 92093, USA

YUAN TAUR

Electrical and Computer Engineering, University of California, San Diego
La Jolla, CA 92093, USA

MARK RODWELL

Electrical and Computer Engineering, University of California, Santa Barbara
Santa Barbara, CA 93106, USA

This paper describes analysis and simulations of Si and III-V Gate-All-Around nanowire MOSFETS assuming ballistic or quasi-ballistic transport. It is found that either channel material can provide the higher saturation current depending on the oxide thickness. For effective oxide thickness above approximately 0.5nm, the higher electron velocity of III-V's outweighs the higher density of states available in the Si device associated with higher effective mass and valley degeneracy and result in higher current for the III-V device. However, materials with higher effective mass and valley degeneracy result in smaller on-resistance in ballistic limit. Depending on the gate oxide capacitance, valley degeneracy may influence the attainable saturation current in a positive or negative way.

Keywords: nanowire; III-V MOSFET

1. Introduction

Recently, scaled MOSFET structures based on nanowires have received intensive research interest since they can provide superior capability in controlling short channel effects (SCE) as compared to conventional bulk structures [1]. For devices with ultra-short gate lengths (comparable to the mean free path of carriers) where these structures are likely to be employed, transport within the devices is expected to be ballistic or quasi-ballistic. In this paper, we compare the ballistic limit performance of III-V and Si nanowire MOSFETs (NWMOSFETs). The simulation reveals interesting tradeoffs between the two types of material. High mass and multi-valley degeneracy may have a positive or negative influence on the saturation current, depending on the effective oxide thickness (EOT) of the device. In all cases, high mass and multi-valley degeneracy results in smaller on-resistance in the ballistic limit at low drain bias than low mass, low valley degeneracy.

2. Numerical Simulation

The structure simulated numerically is a gate-all-around (GAA) MOSFET, as shown schematically in Fig.1. Two types of MOSFETs under study differ only in the core (channel) material: one with III-V core material taken to be $In_{0.53}Ga_{0.47}As$ with effective mass of 0.041, and the other with Si as the core material. The shell insulators are the same for comparison purposes.

Figure 1 Schematic drawing of nanowire transistor.

The simulation employs a self-consistent Schrödinger Poisson solver in cylindrical coordinates to account for the quantum effects. Details of the self consistent solver can be found in [2]. Ballistic current computation is based on an extension [1] of Natori's model [3] where we take explicit account of the energy dependent 1-D Density of States (DOS) while establishing the transverse electrostatic relationship. For completeness of this work, the extended model is briefly recapped below.

In the nanowire case, due to the 2 dimensional confinements along radial and azimuthal directions, the DOS as a function of energy is given by:

$$g(E) = \sum_{\mu,v} \frac{1}{\pi\hbar} \sqrt{\frac{2m^*}{E - E_{\mu,v}}} \cdot u(E - E_{\mu,v}) \tag{1}$$

where m^* is the effective mass along the unconfined direction; $E_{\mu,v}$ is the energy minimum of the subband with indices (μ,v); $u(x)$ is a step function which takes value of 1 when $x \geq 0$ and 0 otherwise. If degeneracy is present, one must multiply the value given in (1) by the degeneracy factor. To compute the ballistic current, both forward going and reverse going current have to be taken into account. The forward going current, which originates from the source, is given by:

$$I_+ = q \sum_{\mu,\nu} \int_{E_{\mu,\nu}} n_+(E)v_+(E)dE = q \sum_{\mu,\nu} \int_{E_{\mu,\nu}} \frac{1}{2}g(E)f(E,E_{f_S})v_+(E)dE \qquad (2)$$

where E_{fS} denotes the Fermi level within the source; $f(E,E_{fS})$ is the Fermi-Dirac distribution function. The factor of *1/2* arises since only the forward going carriers are considered here. Similarly, the reverse going current can be written as:

$$I_- = q \sum_{\mu,\nu} \int_{E_{\mu,\nu}} \frac{1}{2}g(E)f(E,E_{f_S}-qV_{DS})v_-(E)dE \qquad (3)$$

The overall current, is the difference between the forward going and reverse going currents, given by:

$$I = I_+ - I_- = \frac{kTq}{\pi\hbar} \sum_{\mu,\nu} \{\ln[1+\exp(\frac{E_{fS}-E_{\mu,\nu}}{kT})] - \ln[1+\exp(\frac{E_{fS}-qV_{DS}-E_{\mu,\nu}}{kT})]\} \qquad (4)$$

As shown in (4), the relative position of source Fermi level with respect to various subband minima has to be obtained in order to evaluate the ballistic current. This computation is done via solving Poisson's equation along the radial direction (as part of the self consistent computation), where the 1-D DOS is explicitly employed.

In our simulation, Si was approximated as an isotropic material with effective mass of 0.258 ($3*m_t*m_l/(2m_t+m_l)$) and degeneracy factor of 6.

Fig. 2 shows a representative output characteristic for two types of MOSFETs, with EOT equal to 0.55nm for both cases.

Two features are observed here: InGaAs MOSFET delivers higher saturation current, whereas the Si MOSFET exhibits smaller turn-on resistance at low V_{DS}.

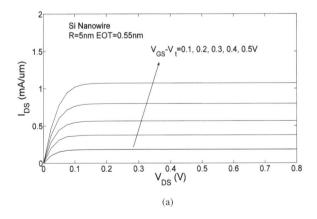

(a)

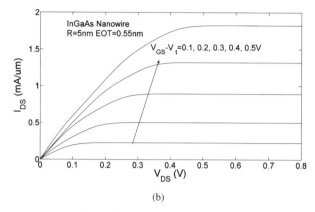

(b)

Figure 2 Computed ballistic output characteristic as function of drain bias for (a) Si nanowire and (b) InGaAs nanowire. The drain Current is normalized to the circumference of the nanowire core.

To understand the higher current in the III-V NWMOSFET, the total charge density and average carrier velocity are shown in Fig. 3 and 4, respectively. It is intuitive that with low density of states (low mass and degeneracy), less charge exists in III-V NWMOSFET. For the structures considered here, at representative bias conditions of $V_{DS}=0.8$ V and $V_{GS}=V_t+0.5V$, the charge density in the Si device is ~3X that of the InGaAs device. The detailed ratio is dependent on the oxide capacitance. However, due to the same reason, the Fermi level within the III-V nanowire follows closely the gate voltage, driving the core material into heavy degeneracy, which induces a sharp increase on average carrier velocity. As shown in Fig. 4, although the ratio of average velocity between InGaAs and Si starts at a factor of ~2.5 ($\sim \sqrt{m^{*}_{Si}/m^{*}_{InGaAs}}$) under non-degenerate conditions, this factor increases to ~8X as the device is driven into heavy degeneracy. For this last case, the increment in carrier velocity outweighs the disadvantage of less charge, thus yielding a higher saturation current for III-V NWMOSFET. A steplike behavior can also be noted in the InGaAs velocity curve, which is due to the discretized eigen energy ladder.

It must be noted that the treatments for both Si and InGaAs devices are based on an assumed parabolic band structure. For III-V materials with strong non-parabolicity (such as InGaAs), this approximation is only strictly valid in a relatively low bias range ($V_{GS}-V_t \leq 0.5V$). With non-parabolicity, the computation task of solving Schrödinger's equation quickly becomes very complex due to the anisotropic band nonparabolicity in the nanowire geometry.

The lower turn on resistance exhibited by the computed characteristics of the above Si device is a result of higher DOS. This result has been calculated for the ballistic limit. For practical devices, when scattering is present, the turn on resistance will also be influenced by the low field mobility of the material. The advantage in turn-on resistance of the Si device may diminish as a result of its lower mobility than in InGaAs.

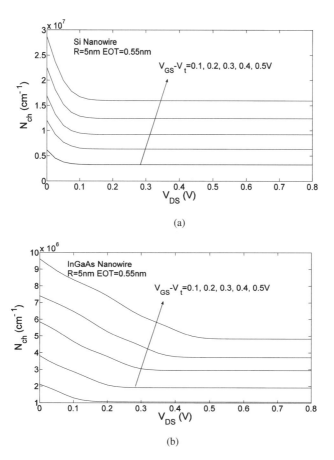

(a)

(b)

Figure 3 Computed ballistic line charge density as function of drain bias for (a) Si nanowire and (b) InGaAs nanowire.

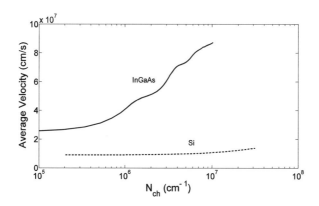

Figure 4 Average injection velocities of Si and InGaAs at the virtual source.

3. Analysis and Discussion

To further understand the tradeoffs associated with effective mass and valley degeneracy under different EOTs', we analyze the structure at T=0K, to avoid involved mathematics.

We start by defining a capacitance network along the radial direction, so as to obtain an analytical relationship between applied gate bias and the source Fermi level at the interface. Under the scenario that most carriers reside in one subband, a perturbation in gate voltage δV_G can be expressed as:

$$\delta V_G = \delta\phi_{OX} + \frac{1}{q}\delta(E_{fs} - E_1) + \frac{1}{q}\delta(E_1 - E_{C0}) \tag{5}$$

where $\delta\Phi_{OX}$ is the potential drop across the insulator (oxide), and E_1 is the first subband minima, E_{C0} is the bottom of the conduction band at the interface. The overall input capacitance is defined as:

$$C_{in} = q(\frac{dV_G}{dN})^{-1} = q(\frac{d\phi_{OX}}{dN} + \frac{1}{q}\frac{d(E_{fs} - E_1)}{dN} + \frac{1}{q}\frac{d(E_1 - E_{C0})}{dN})^{-1} \tag{6}$$

The first term within the brackets on the right hand side of (6) is clearly $1/C_{ox}$, while we define the second term to be the DOS capacitance C_{DOS} and the third term to be quantum well capacitance C_{qw}. Thus, the input capacitance can be broken up into a network as shown in Fig. 5.

Figure 5 Schematic of the capacitance network.

It can be shown [1] that the forward going line charge density of a core/shell nanowire structure, at T=0K, can be written as:

$$N = \frac{g_d}{2\pi\hbar}\int_{E_1}^{E_{fs}}\sqrt{\frac{2m^*}{E - E_1}}dE \tag{7}$$

where g_d is the valley degeneracy. The C_{DOS} term can therefore be simplified into a form resembling the 1-D DOS:

$$C_{DOS} = (q^2 g_d / \pi\hbar)\sqrt{m^*/2(E_{fs} - E_1)} \tag{8}$$

C_{qw} is related to the width of the wave function, and is generally slowly varying with gate bias [4]. Therefore, C_{qw} and C_{ox} may be combined to form an effective oxide capacitance C_{ox}'.

At T=0K, from Eq. (4), the ballistic current can be written as

$$I = g_d q(E_{fs} - E_1)/\pi\hbar . \tag{9}$$

where $(E_f\text{-}E_1)$ can be solved via the capacitance network through the relation

$$\frac{1}{q}d(E_{fS} - E_1) = d(V_G - V_{th}) \cdot \frac{C_{ox}{}'}{C_{ox}{}' + C_{DOS}}$$

(10)

The ballistic current can thus be expressed as:

$$I = \frac{q^2 g_d}{\pi\hbar}[(V_G - V_{th}) - \frac{m^* q^3 g_d{}^2}{\pi^2\hbar^2 C_{ox}{}'^2}(\sqrt{1 + \frac{2\hbar^2\pi^2 C_{ox}{}'^2(V_G - V_{th})}{q^3 g_d{}^2 m^*}} - 1)]$$

(11)

Plotted in Fig. 6 is the ballistic current expressed by (11) with various effective mass and degeneracy combinations. In general, the ballistic current decreases with effective mass with a given EOT value. Interestingly, depending on the actual effective oxide thickness, the ballistic current may or may not benefit from high effective mass and high valley degeneracy. This crossover behavior can be understood by considering extreme conditions:

(1) $C_{ox}{}' >> C_{DOS}$: the current approaches a constant value of $I = (g_d q^2/\pi\hbar)(V_G\text{-}V_{th})$, which is proportional to the degeneracy factor. This asymptotic behavior implies that the current becomes (almost) independent of EOT at low effective mass.

(2) $C_{ox}{}' << C_{DOS}$: by expanding the square root term in (11) to the second order, we obtain another asymptotic expression, $I = \pi\hbar C_{ox}{}'^2(V_G - V_{th})^2/(2m^* g_d q)$. This indicates that the current decreases with increased effective mass and valley degeneracy. This effect shows up at the high effective mass and with high EOT value (see EOT=3nm curve, where dotted line representing material with higher valley degeneracy fall below the solid line representing lower valley degeneracy at high effective mass).

These two opposite asymptotic trends thus predict a crossover between low and high valley degeneracy as the effective mass increases. The position where the crossover occurs is dependent on the relative magnitude of the (effective) oxide capacitance and the DOS capacitance. Therefore, high mass or high valley degeneracy may have a positive or

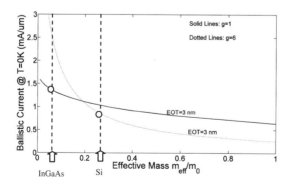

Figure 6 Computed ballistic saturation current at T=0K with one subband only, with gate overdrive of 0.5V. The EOT considered here includes both oxide thickness and finite wave function depth. Solid curves are for single valley situation, dotted curves are for valley degeneracy of 6. Current is normalized to circumference of the nanowire core. m_{eff}/m_0 values appropriate to InGaAs and Si are shown with symbol (o).

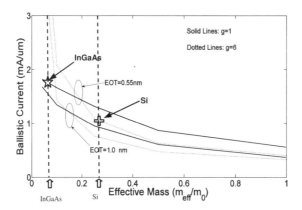

Figure 7 Numerical simulation of ballistic saturation current at room temperature with gate overdrive of 0.5V. The EOT considered here includes ONLY the oxide thickness. Solid curves are for single valley situation, dotted curves are for valley degeneracy of 6. Current is normalized to circumference of nanowire core.

negative effect on the device performance in the ballistic limit. The same trend is confirmed with the full numerical simulation that takes into account finite temperature and multiple subbands, as plotted in Fig. 7. It is worthwhile to notice that for EOT= 0.55nm, InGaAs (symbol star) exhibits higher ballistic current than Si (symbol cross).

4. Conclusions

In conclusion, ballistic III-V NWMOSFETs may outperform Si NWMOSFET in terms of saturation current depending on the available equivalent insulator thickness. Large mass and greater valley degeneracy always result in a smaller turn on resistance at low drain bias in the ballistic limit. Valley degeneracy may have positive or negative influence on saturation current depending on the EOT.

5. Acknowledgement

This work has been partly supported by the SRC Non-classical CMOS Research Center (2006-VC-1437) and partly by the National Science Foundation.

References

1. B. Yu, L. Wang, Y. Yuan, Y, Taur, P. Asbeck, "Scaling of nanowire transistors", accepted by Trans. of Elec. Dev.
2. L. Wang, D. Wang, P. Asbeck, "A numerical Schrödinger-Poisson solver for radially symmetric nanowire core-shell structures", Solid State Electronics, vol. 50, pp 1732-1739
3. K. Natori, "Ballistic metal-oxide-semiconductor field effect transistor", J. Appl. Phys. vol. 76, pp. 4879-4890

International Journal of High Speed Electronics and Systems
Vol. 19, No. 1 (2009) 23–31
© World Scientific Publishing Company

A ROOM TEMPERATURE BALLISTIC DEFLECTION TRANSISTOR FOR HIGH PERFORMANCE APPLICATIONS

QUENTIN DIDUCK*

*Electrical and Computer Engineering, Cornell University, 426 Phillips Hall
Ithaca, New York 14853, USA
diduck@ieee.org*

HIROSHI IRIE[†]

*Electrical and Computer Engineering, University of Rochester, Hopeman 342,
Rochester, New York 14850, USA
irie@ece.rochester.edu*

MARTIN MARGALA[‡]

*Electrical and Computer Engineering, University of Massachusetts Lowell,
Lowell, Massachusetts, 01854, USA
martin_margala@uml.edu*

The Ballistic Deflection Transistor (BDT) is a novel device that is based upon an electron steering and a ballistic deflection effect. Composed of an InGaAs-InAlAs heterostructure on an InP substrate, this material system provides a large mean free path and high mobility to support ballistic transport at room temperature. The planar nature of the device enables a two step lithography process, as well, implies a very low capacitance design. This transistor is unique in that no doping junction or barrier structure is employed. Rather, the transistor utilizes a two-dimensional electron gas (2DEG) to achieve ballistic electron transport in a gated microstructure, combined with asymmetric geometrical deflection. Motivated by reduced transit times, the structure can be operated such that current never stops flowing, but rather is only directed toward one of two output drain terminals. The BDT is unique in that it possesses both a positive and negative transconductance region. Experimental measurements have indicated that the transconductance of the device increases with applied drain-source voltage. DC measurements of prototype devices have verified small signal voltage gains of over 150, with transconductance values from 45 to 130 mS/mm depending upon geometry and bias. Gate-channel separation is currently 80nm, and allows for higher transconductance through scaling. The six terminal device enables a normally differential mode of operation, and provides two drain outputs. These outputs, depending on gate bias, are either complementary or non-complementary. This facilitates a wide variety of circuit design techniques. Given the ultralow capacitive design, initial estimates of f_t, for the device fabricated with a 430nm gate width, are over a THz.

* Quentin Diduck is with the Electrical and Computer Engineering Department, Cornell University, Ithaca, NY 14850 USA (e-mail: diduck@ieee.org).
† The Late Marc Feldman was with the Electrical and Computer Engineering Department, University of Rochester, Rochester, NY 14627 USA.
‡ Martin Margala, is with Electrical and Computer Engineering Department, University of Massachusetts Lowell, Lowell, MA 01854 USA (e-mail: martin_margala@uml.edu)

1. Introduction

Since the introduction of the Bipolar Junction Transistor and Field Effect Transistor, little progress toward novel transistor designs has been achieved. Evolutionary changes in size and material compositions have provided better devices, however these are just improvements to a fundamentally similar structure. Recent work on Y-branch structures and ballistic transport effects has opened up the possibility of new triode type devices.[1,6,11] The scaling limits of Field Effect and Bipolar devices are quickly being reached with current leakage power essentially limiting further development.[5] The need for a novel structure that provides higher performance in the same or less area is upon us. In addition to scaling transistors, to reduce capacitance and therefore improve frequency response, a variety of material systems are been utilized. Higher mobility semiconductors such as GaAs, InP and InSb have been used to improve the frequency response of a transistor by reducing the electron transit time through the device and providing a high current drive to achieve a large f_t. The novel structure presented here is intended to provide a high speed response by supporting a continuous current flow, and by controlling its direction. It is this approach that is believed will greatly reduce transit times and lead to a large f_t.

2. Background

The physical structure of the BDT implies several different physical effects participate in the operation of the device. The lateral field effect that is responsible for steering the electrons has been studied by Reitzenstein et al.[11] This work was conducted at low temperatures and indicated that ballistic transport between two lateral gates induced a steering effect that can direct current into one of two drains. The results of this work indicated that Y-branch style devices have a drain bias dependence on the overall transconductance, at least when operated in the ballistic regime.

In our work, we strengthened the electric field in the device by adding a third drain terminal, combined with an asymmetric scattering object. This concept is based upon the research by Song et al. and his ballistic rectifier experiments.[3,8,9] The ballistic rectifier structure produces a non-linear response through the deflection of electrons. In the ballistic rectifier experiments, a triangular artificial scattering mechanism is introduced into the center of a cross shaped structure. By applying an AC voltage across this triangular structure, electrons are preferably directed to the lower channel. The effect is shown clearly to be related to ballistic transport, as the magnitude of the effect depends on temperature and geometry scale. This work counters the generally accepted theory describing ballistic transport, which is the linear Landauer-Büttiker theory.[7,10] Thus, a non-linear extension is needed to explain ballistic rectifier experiments. While many modifications have been suggested, no clear theory has been established.

3. Theory of Operation

The general operation of the BDT is implicit with its structure. In many ways, the BDT can be thought of as a semiconductor implementation of a two dimensional Cathode Ray Tube (CRT). In essence, the lateral gates of the channel direct electrons to either the left or right drain (see Figure 1). A bias point at the center of the device aids in adjusting the gain. However, an additional bias potential is required on at least one drain in order for the device to function as a transistor. Differential input voltages on the gates are used to steer electrons into one channel or the other. Over steering the current, results in a pinch-off effect. Increasing the drain voltage or the center bias voltage increases the transconductance of the device by increasing the electron energy/velocity.

The central deflector aids in directing current into one channel or the other and reduces current lost to the center bias port. Devices without this deflective structure have been fabricated for comparison. The center currents on these devices were considerably higher than either of the other drain ports even though all the port sizes were the same. It is expected that the deflective structure will become more significant with further device scaling. As we reduce the scale, a larger percentage of electrons will interact with the artificial scattering object ballistically. This should produce an effect similar to the ballistic bridge rectifier. We can consider the region above the gates as approximately equivalent to two ballistic bridge rectifier structures super imposed (rectifying to the left and right drain). To simplify the analysis, we first consider the simplest case, where we have no bias on the right or left drain, but only the center drain, which is essentially identical to the bridge rectifier setup. Treating it as such, from Song we consider for small currents, electrons that are directed toward the right drain induce the following:

$$V_{RL} = -\frac{3\pi h^2}{4e^3 N_{SCD} E_F N_{RD}} I_{SR}^2. \qquad (1)$$

An electron directed toward the left drain correspondingly induces:

$$V_{RL} = -\frac{3\pi h^2}{4e^3 N_{SCD} E_F N_{LD}} I_{SL}^2, \qquad (2)$$

where E_F is the Fermi Energy I_{SR} is the current in the right channel, I_{SL} is the current in the left channel, and N is the number of conducting modes (LD for Left Drain, RD for Right Drain, and SCD for the Center Drain).[9]

If we sum V_{RL} and V_{LR} we get a net potential between the two drains. With both gates at an equal voltage, V_{RL} and V_{LR} are equal, however as the gates change the current in each channel, a potential difference increases non-linearly. The non-linear change in potential is due to electrons being deflected away from the center drain port by the artificial scattering mechanism. While the calculations here assume an equal number of conducting modes at the source and the center drain, in the fabricated device this is only true when only one of the channels is fully conducting. Thus for an exacting equation, the individual transmission probabilities for each port need to be considered. Nonetheless, we see that the triangular deflective structure will further enhance the non-linearity of the device as we scale and a larger percentage of the electrons will interact ballistically, enhancing the transconductance of the device.

The gate control over the channel can be attributed to two effects, classic channel pinch-off due to field effect, and an electron steering effect. The device can be operated as a field effect device by applying the same voltages to each gate. However, the more interesting behavior occurs when different voltages are applied. In this case, we obtain an electron steering effect, and the coupling of this lateral structure is stronger than typical top gate designs. This additional coupling is attributed to the reduction of charge screening effects.[2] The steering effect implies the flowing electrons encounter a lateral field directing them towards one channel or the other, depending on gate bias. Pinch-off can be observed when a strong bias depletes the channel completely. Since there are both pinch off and steering effects, the transconductance of the device can be either positive or negative depending upon the gate voltage.

The f_t performance of this device can be estimated using the approximation:

$$f_t = g_m \Big/ \left(2\pi * C_g\right),\qquad(3)$$

where g_m is the transconductance and C_g is the gate capacitance of the BDT. We estimated the gate capacitance including fringe fields by assuming the gate 2DEG region is coupled to an infinite conducting plane at half the distance between the gate-channel region, and assume an equivalent capacitance from the channel to the infinite plane. An effective dielectric constant of 6.57 was derived using a conservative microstrip approximation.[4] The capacitance equation for a rectangle to an infinite plane is:

$$C = \left(\left(\frac{s_0 s_r A}{d}\right)^{\eta} + \left(C_0 \frac{s_r + 1}{2}\right)^{\eta}\right)^{\frac{1}{\eta}},\qquad(4)$$

$$\text{where } C_0 = 0.9\varepsilon_0 \sqrt{8\pi A},\qquad(5)$$

$$\text{and } \eta = 1.114,$$

Using the channel-gate separation of 80nm, a 430nm gate width and a 2DEG thickness of 9 nm, the estimation yields a capacitance of approximately 7.02 aF. Using eq. 3 with a g_m of $4.5\text{x}10^{-5}$ S, this yields a theoretical maximum f_t of approximately 1.02 THz. The planner nature of the BDT facilitates the low capacitance design. While great care has been taken to calculate this capacitance value, only direct f_t measurements can determine the true capabilities of this device. The differential input requirement poses certain technical challenges that have to be over come in order to perform a direct f_t measurement.

4. Experiments and Results

Several devices of different geometries were fabricated on separate runs. The substrate was fabricated using Molecular Beam Epitaxy, and consists of lattice matched layers of InGaAs and InAlAs on an InP substrate. Device fabrication consists of only two lithography steps. First, a suitable hard-mask is patterned using electron beam lithography. This mask is then used in the etching of the substrate via an inductively coupled reactive ion etcher. Contacts are formed using a standard Ni-Ge-Au stack that was annealed at 400 degrees Centigrade until ohmic. In Figure 1, a scanning electron microscope image of a typical BDT structures is shown. The dark areas including the triangle region is removed material from the 2DEG structure. Several IV curve measurements were performed on a six probe station using an Agilent 4156C Parameter Analyzer at room temperature.

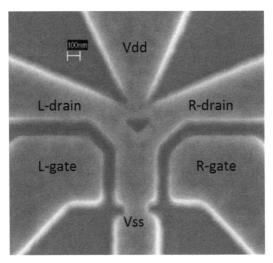

Figure 1. Pictured is a BDT with 500nm gates (including the angled region) and 80nm gate-channel spacing, the top-left and top right ports are drain ports, bottom left and right ports are gates, top port is a vdd bias port that controls gain, and the bottom port is the source. Dark regions indicate removed material.

From the IV curve measurements several interesting effects were observed. In Figure 2, we see the current response for a series of drain voltages, as a function of gate bias. The gates were biased push-pull with the left gate used as the reference in the charts (as an example of a push pull bias, when the left gate is at -0.1V the right gate will be at +0.1V). There are two significant results of this measurement. First, we observe that the current first increases as a function of gate voltage then decreases. Second, we notice that the drain current increases as a function of applied bias as well as the transconductance. At the peak between the steering region and the pinch-off region, we have maximum conductivity. The right drain (which is not shown) has the identical response, but mirrored about the center axis. Subtle differences in amplitude occur due to process variation, though some devices have been measured with near identical left and right drain response. This positive and negative transconductance region characteristic, enables circuits that are inverting and non-inverting, depending only on gate offset voltage.

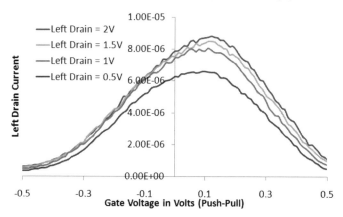

Figure 2. IV characteristic of the left drain port as a function of push-pull gate voltage (in reference to the left gate), an increase in drain voltage increases the transconductance, near linearly for voltage beyond 1V.

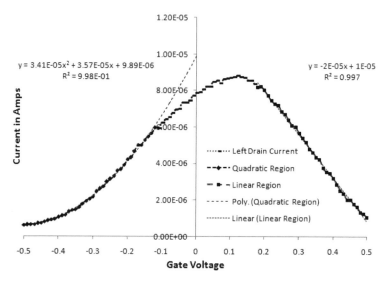

Figure 3. Quadratic and linear curve fitting of the current response as a function of gate voltage.

We conducted a numerical curve fit shown in Figure 3, the results suggest that there is a quadratic and linear current region. The quadratic region suggests that the channel is being modulated by a depletion region in a similar fashion as a field effect device whereby as the gate voltage becomes more positive, a conducting channel opens. The linear region appears to form as a result of the channel being closed off by a depletion region induced by the other gate, even though locally the channel is being enhanced by the positive gate. As suggested by K. Hieke et al., it appears that the negative voltage dominates the modulation of the channel.[12] The region with a gate voltage between -0.1 Volts and 0.2 Volt, for the graph in Figure 3, is the transition region between the quadratic channel opening and the linear pinch-off of the channel.

The positive and negative slopes of the drain current as a function of gate voltage necessitates the use of two graphs to plot drain current versus drain-source voltage. In Figure 4a the negative gate voltages are shown, and the positive gate voltages are shown in Figure 4b. We note that for the negative gate regions, as expected from the previous graphs, the current rises with increasing voltage. The converse is true for positive gate voltages. The quadratic region would require a different amplifier design technique than the linear region. As well the transition region requires more study before amplifiers operating in this region can be well designed.

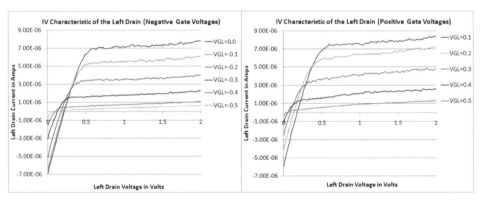

Figure 4. a) IV Current response as a function of drain voltage for multiple NEGATIVE push pull gate voltages (shown in reference to the left gate). Note that for negative gate voltages the current increases as gate voltage increases. b) IV Current response as a function of drain voltage for multiple POSITIVE push pull gate voltages (shown in reference to the left gate). Note that for positive gate voltages the current decreases as gate voltage increases.

In all the figures only the left drain response is shown, but it is important to consider that this device has a mirrored response on the right drain terminal. Thus, it is possible to design differential circuits by using only one transistor. This can provide a significant benefit for communication circuit design, as well as potentially enable higher signaling rates by utilizing differential signaling throughout the circuit design.

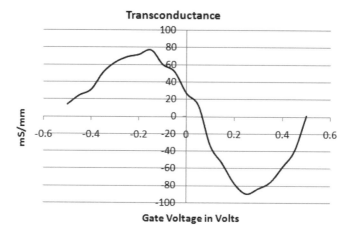

Figure 5. Transconductance of the BDT shown in Figure 1 as a function of push-pull gate voltage with a 2V drain bias applied.

In Figure 5 the transconductance as a function of gate voltage is plotted for a 2V bias. Drain bias affects transconductance performance, and this suggests that higher electric fields will produce devices with higher performance. Device scaling will enable designs that have higher electric fields as well as will ensure that a larger percentage of the electrons are being transported ballistically through the device rather than by drift.

Experimental measurements have indicated that the transconductance of the device increases with applied drain-source voltage. DC measurements of prototype devices have verified small signal voltage gains of over 150 V/V, with transconductance values from 45 to 130 mS/mm depending upon geometry and bias. Gate-channel separation is currently 80nm, and allows for higher transconductance through scaling.

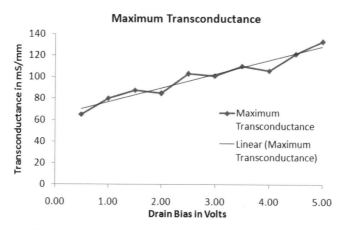

Figure 6. Change in transconductance as a function of applied drain bias and a linear fit.

5. Conclusion

In conclusion, we have presented a novel transistor structure that is suitable for high speed applications. The planar design and mathematical estimates suggest the BDT design has a very low gate capacitance that can potentially enable THz operation. The device has two drain terminals that can be configured to operate differentially, facilitating differential circuit design using only one transistor. It has been demonstrated that this devices' transconductance increases with applied drain voltage. This supports the notion that as the device is scaled, transconductance will improve not only by reducing gate-channel distance, but also by shortening the channel length. The highly configurable nature of the Ballistic Deflection Transistor makes it suitable for a wide variety of applications.

Acknowledgments

The authors would like to thank Daniel Purdy for his comments and contributions to the work as well as for providing the initial support of this project under ONR award N000140510052. This research is also funded in part by NSF under award number 0609140 and by the AFOSR under award number FA9550-07-1-0032.

References

1. J. G. Cardoso, *Acta. Phys. Pol. B* **32**, 29 (2001).
2. Linda Geppert, IEEE Spectrum, pp 20-21 Jan 2003
3. A.M. Song, P. Omling, L. Samuelson, W. Seifert, I. Shorubalko and H. Zirath, Jpn. J. Appl. Phys. **40** L 909 (2001).
4. Robert E. Collin, *Foundations for Microwave Engineering* (McGraw-Hill, New York, 1992.)
5. W. Haensch, E. J. Nowak, R. H. Dennard, P. M. Solomon, A. Bryant, O. H. Dokumaci, A. Kumar, X. Wang, J. B. Johnson, M. V. Fischetti, IBM Journal of Research and Development, Vol **50** , Issue **4/5** (July 2006) Pages: 339 – 361
6. J.-O.J. Wesström, Phys. Rev. Lett. **82**, 2564 (1999).
7. R. Landauer, Philos. Mag. **21**, 863 (1970).
8. A. M. Song, A. Lorke, A. Kriele, and J. P. Kotthaus, Phys. Rev. Lett. **80**, 3831 - 3834 (1998)
9. A. M. Song, Vol **59**, No **15**, Physical Review B,15 Apr 1999.
10. M. Büttiker, Phys. Rev. Lett. **57**, pp. 1761, 1986.
11. S. Reitzenstein, L.Worschech, P. Hartmann, M. Kamp, and A. Forchel, Phys. Rev. Lett. **89**, 226804 (2002).
12. K. Hieke, J. Wesstrom, T. Palm, B. Stalnacke, and B. Stoltz, Vol **42**, No **7-8**, Solid-State Electronics pp 1115-1119, 1998

International Journal of High Speed Electronics and Systems
Vol. 19, No. 1 (2009) 33–53
© World Scientific Publishing Company

EMISSION AND INTENSITY MODULATION OF TERAHERTZ ELECTROMAGNETIC RADIATION UTILIZING 2-DIMENSIONAL PLASMONS IN DUAL-GRATING-GATE HEMT'S

TAIICHI OTSUJI

*Research Institute of Electrical Communication, Tohoku University, 2-1-1 Katahira,
Sendai, Miyagi 980-8577, Japan
otsuji@riec.tohoku.ac.jp*

TAKUYA NISHIMURA

*Research Institute of Electrical Communication, Tohoku University, 2-1-1 Katahira,
Sendai, Miyagi 980-8577, Japan*

YUKI TSUDA

*Research Institute of Electrical Communication, Tohoku University, 2-1-1 Katahira,
Sendai, Miyagi 980-8577, Japan*

YAHYA MOUBARAK MEZIANI

*Research Institute of Electrical Communication, Tohoku University, 2-1-1 Katahira,
Sendai, Miyagi 980-8577, Japan*

TETSUYA SUEMITSU

*Research Institute of Electrical Communication, Tohoku University, 2-1-1 Katahira,
Sendai, Miyagi 980-8577, Japan*

EIICHI SANO

*Research Center for Integrated Quantum Electronics, Hokkaido University, N13W8,
Sapporo, Hokkaido 060-8628, Japan*

Two dimensional plasmons in submicron transistors have attracted much attention due to their nature of promoting emission/detection of electromagnetic radiation in the terahertz range. We have recently proposed and fabricated a highly efficient, broadband plasmon-resonant terahertz emitter. The device incorporates doubly interdigitated grating gates and a vertical cavity into a high electron mobility transistor. The device operates in various modes: (1) DC-current-driven self oscillation, (2) CW-laser excited terahertz emission, (3) two-photon injection-locked difference-frequency terahertz emission, and (4) impulsive laser excited terahertz emission. Furthermore, the device can operate in completely different functionalities including ultrahigh-speed intensity modulation for terahertz carrier waves. This paper reviews recent advances on plasma wave devices.

Keywords: Terahertz; plasmon; resonance; emitter; instability; injection locking; intensity modulator

1. Introduction

"Terahertz" is an unexplored frequency band in the sense that there is no commercially available microelectronic device that can generate, detect, or manipulate electromagnetic waves over the entire terahertz frequency band[1]. In the last decade, then therefore, development of compact, tunable and coherent sources operating at terahertz frequencies has been one of the hottest issues of the modern terahertz (THz) electronics[1]. Two dimensional (2D) plasmons in submicron transistors have attracted much attention due to their nature of promoting emission of electro-magnetic radiation in the terahertz range[2-5]. We have recently proposed and fabricated a highly efficient, broadband plasmon-resonant terahertz emitter/photomixer device[6-11]. The device incorporates doubly interdigitated grating gates and a vertical cavity into a semiconductor heterojunction high electron mobility transistor (HEMT) so that structure-dependent highly dispersive plasmonic systems can be configured in submicron-to-nanometric scaled artificial dimensions to perform emission, detection, and moreover higher functional signal processing like intensity modulation as well as frequency multiplication in an exploring terahertz frequency region. Test samples are fabricated using InGaP/InGaAs/GaAs material systems, succeeding in the first observation of stimulated emission of terahertz radiation at room temperature. This article reviews recent advances in novel plasmonic nanotransistors for emission and intensity modulation of terahertz electromagnetic waves. and presents a new result on photomixed injection locked oscillation.

2. Plasmon-Resonant Terahertz Emitter

2.1. *Device structure and operation principle*

Figure 1 illustrates the cross section of the plasmon-resonant emitter. The device structure is based on a high electron mobility transistor (HEMT) and incorporates (i) doubly interdigitated grating gates (G1 and G2) that periodically localize the 2D plasmon in stripes on the order of 100 nm with a micron-to-submicron interval and (ii) a vertical cavity structure in between the top grating plane and a terahertz mirror at the backside. The structure (i) works as a terahertz antenna[12] and (ii) works as an amplifier. The terahertz mirror is to be a transparent metal like indium titanium oxide (ITO) when the device works in an optical excitation mode so as to excite the plasmons by optical two-photon irradiation from outside the back surface[6].

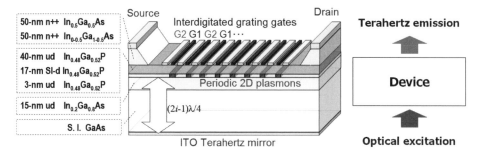

Fig. 1. Device cross section for typical GaAs-based heterostructure material systems. k: the wave vectors of irradiated photons, E_x: the electric field (linear polarization), k_{THz}: the wave vector of electromagnetic radiation.

Suppose that the grating gates have geometry with 300-nm G1 fingers and 100-nm G2 fingers to be aligned alternately with a space of 100 nm and that an appropriately high 2D electronic charge ($\sim 10^{12}\mathrm{cm}^{-2}$) is induced in the plasmon cavities under G1 while the regions under G2 are weakly charged ($10^{10}\sim 10^{11}\mathrm{cm}^{-2}$). Figure 2 depicts a numerically simulated typical 2D electron density/velocity distributions based on a self-consistent drift-diffusion Poisson equations. A standard DC drain-to-source bias V_{DS} of 50 mV/(grating period), and the gate biases V_{G1} and V_{G2} of V_{th} + 2.2 V and V_{th} + 0.2 V are assumed where V_{th} is the threshold voltage. As is seen in Fig. 2, a strong electric field (1~10 kV/cm) arises at the plasmon cavity boundaries[11]. When the DC drain-to-source bias V_{DS} is applied, 2D electrons are accelerated to produce a constant drain-to-source current I_{DS}. Due to such a distributed plasmonic cavity systems with periodic 2D electron-density modulation, the DC current flow may excite the plasma waves in each plasmon cavity. As shown in Fig. 3, asymmetric cavity boundaries make plasma-wave reflections as well as abrupt change in the density and the drift velocity of electrons, which may cause the current-driven plasmon instability[3, 13, 14, 15] leading to excitation of coherent resonant plasmons. Thermally excited hot electrons also may excite incoherent plasmons[16-20]. The grating gates act also as terahertz antenna that converts non-radiative longitudinal plasmon modes to radiative transverse electromagnetic modes[6].

When the device is photoexcited by laser irradiation, photoelectrons are predominantly generated in the weakly-charged regions with many unoccupied electronic states under G2 and then are injected to the plasmon cavities under G1. Thanks to a specific drain-to-source bias promoting a uniform slope along the source-to-drain direction on the energy band in the regions under G2, photoelectrons under G2 are unidirectionally injected to one side of the adjacent plasmon cavity. This may also excite the plasmons under an asymmetric cavity boundary[3, 4, 21]. It is noted that the laser irradiation may excite the plasmon not only in the regions under G1 but also in the regions under G2 if the cavity size and carrier density of the regions under G2 also satisfies the resonant conditions.

Once the terahertz electromagnetic waves are produced from the seed of plasma waves, downward-propagating electromagnetic waves are reflected at the mirror back to the plasmon region so that the reflected waves can directly excite the plasmon again according to the Drude optical conductivity and intersubband transition process[6]. When the plasmon resonant frequency satisfies the standing-wave condition of the vertical cavity, the terahertz electromagnetic radiation will reinforce the plasmon resonance in a recursive manner. Therefore, the vertical cavity may work as an amplifier if the gain exceeds the cavity loss. The quality factor of the vertical cavity is relatively low as is

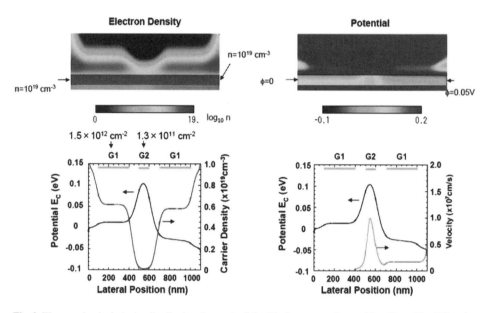

Fig. 2. Electron density/velocity distributions in a unit of the 2D plasmon grating cavities. $V_{DS} = 50$ mV/(grating period), $V_{G1} = V_{th} + 2.2$ V, $V_{G2} = V_{th} + 0.2$ V. (after Ref. 11.)

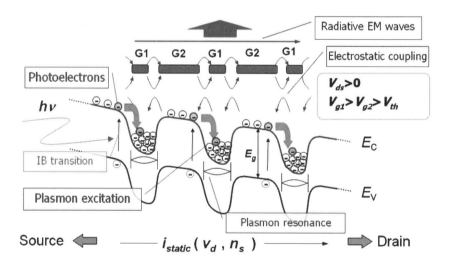

Fig. 3. Schematic band diagram and operation mechanism. (after Ref. 11.)

simulated in Refs. 6, 22 since the 2D plasmon grating plane of one side of the vertical cavity boundary must have a certain transmittance for emission of radiation. Thus, the cavity serves a broadband character.

2.2. *Characteristic parameters and design scheme*

Field emission properties of the dual-grating-gate plasmon-resonant emitters are characterized by the structure dependent key parameters shown in Fig. 4[6, 22]. The primary parameter that initiates the plasmon resonance is ω_{p2} which is the plasma frequency, i.e. plasmon characteristic frequency, of the periodically confined *gated* plasmon cavities. Each cavity is connected by the connecting portion whose carrier density must be controlled to be far apart from that in the plasmon cavity to make a good plasmon confinement. Thus, this connecting portion has its characteristic frequency ω_{p3}. The grating gate has also its own plasma frequency ω_{p1}. Note that ω_{p2} and ω_{p3} for the *gated* plasmon cavities and connecting portions obey the linear dispersion law while ω_{p1} for the *ungated* gate gratings is proportional to the square-root of wave vector[3, 6, 13]. All the three parameters are mainly determined by their cavity length: W, the distance between layers: d, and carrier density, and perturbed by their periodicity: a, or the filling parameter: $f = W/a$ [13]. The final parameter, denoted by ω_L, corresponds to the vertical cavity resonance.

According to the operating frequency band, the grating geometry (single plasmon cavity length and periodicity) is designed to be fixed and ω_{p1} and ω_{p3} as well as ω_L are optimally designed. For an actual device operation, ω_{p2} is a given parameter, which is first tuned by the gate bias at a specific value to obtain a desired resonance frequency. As a fundamental design criterion to obtain high quantum efficiency, ω_{p1} and ω_L values are to be matched to ω_{p2} value while ω_{p3} is far depart from them. Once the device dimensions and material systems are designed, ω_{p1} and ω_L become fixed parameters. ω_{p3} for the connecting portion, on the other hand, is controllable (by V_{g2}) so that one can set it at far higher or lower than ω_{p2} by making the connecting portion to be metallic or dielectric. When the grating gate is made with metals like standard HEMT's, ω_{p1} becomes higher by orders of magnitude than ω_{p2}, resulting in degrading the emission power/efficiency[6, 21]. To prevent it, a semiconducting material, in particular, 2DEG grating gate made from the upper deck of a double-decked HEMT is superior[11, 23, 24], which is demonstrated in Sec. 5.

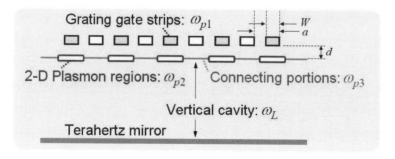

Fig. 4. Characteristic frequencies. (after Ref. 6.)

The plasma wave behavior of the 2D electron systems (2DES) is described by the extended Dyakonov-Shur model[3, 6]. Under the gradual channel approximation, the local electron density n and velocity v of the plasma fluid are formulated by the hydrodynamic equations:

$$m_e \left(\frac{\partial v}{\partial t} + (v \cdot \nabla) v \right) = -e\nabla U - m_e \frac{v}{\tau} \ , \tag{1}$$

$$\frac{\partial n}{\partial t} + \nabla(nv) = \frac{\partial U}{\partial t} + \nabla(Uv) = 0 \ , \tag{2}$$

where m_e the electron effective mass, e the electronic charge, U the gate-to-channel potential, τ the plasmon relaxation time. Their time-evolved response to the terahertz excitation is numerically analyzed using the finite differential time-domain (FDTD) method. The plasma waves themselves are the coherent electronic polarization so that they should produce local displacement AC current. Thus, it is input to the Maxwell's FDTD simulator as a current source to analyze the electromagnetic field dynamics.

Figure 5 shows typical instantaneous cross-sectional distribution of the electric-field intensity along the x (source to drain) direction under a constant sinusoidal plasmon excitation at (a) a tuned frequency of 3.4 THz and (b) a detuned frequency of 5.1 THz[6]. The device model is based on the HEMT shown in Fig. 1 accommodating nine periods of the dual grating gates. The gate finger lengths, L_{G1} and L_{G2}, of 200 and 900 nm with 100-nm spacing are assumed. The characteristic frequencies ω_{p1} and ω_L are set at 3.4 THz. The primary parameter ω_{p2} are set at the excitation frequency (3.4 THz for (a) and 5.1 THz for (b)). All the plasmon cavities are excited in phase. One can see in Fig. 5 (a) an in phase oscillation between outside air (upper portion) and inside the cavity since the vertical cavity length is set at the quarter wavelength of the fundamental mode. The white colored area shows very high intensity of over the range. Under a detuned condition of 5.1-THz excitation, on the other hand, antiphase oscillation is seen as is expected. The radiation power is almost remained at the level of that for the tuned condition. For both cases, the periodic longitudinal polarization in the plasmon grating plane is satisfactorily

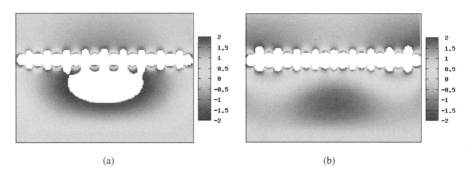

(a)   (b)

Fig. 5. Simulated instantaneous electric field (E_x) distributions under a constant sinusoidal plasmon excitation (a) at 3.4 THz and (b) 5.1 THz. Intensity scaled on the indicator is in arbitrary unit. (after Ref. 6.)

converted to the transverse monotonic electric field in the outside the device (upper portion in Fig. 5). The results clearly show the standing wave oscillation inside the cavity and forward propagating quasi-transverse electromagnetic (TEM) waves outside in air.

In order to examine how the double gate grating and vertical cavity structures contribute to the field emission properties, artificial structures without double gate grating and/or terahertz mirror are prepared for, and compared their impulse responses to that of original structure by using Maxwell's FDTD simulator[6]. All the characteristic parameters were fixed at the nominal values ($\omega_{p1}= \omega_{p2}= \omega_{L}=3.4$ THz). Each plasmon cavity was excited with an impulsive current source simultaneously. Simulated temporal responses of the electric field (x component) at the central two points (4 μm beyond the gate surface and 4 μm beneath the plasmon surface) were Fourier transformed to obtain entire frequency spectra.

Figure 6 plots the results[6]. For the structures without gate gratings, neither *gated* plasmon modes nor the Smith-Purcell effect is produced resulting in no obvious field enhancement over the frequency range; a small dip below 1THz is an unphysical error caused in numerical process. The vertical cavity makes a resonance property and weakly enhances the radiation in narrow bands around the fundamental and second harmonic frequencies.

Incorporating the double gate grating, on the contrary, produces extraordinary electromagnetic transmission; the electric field intensity drastically enhances over a broadband range. As a result, mode-conversion gain, from non-radiative plasmon mode to radiative mode, of up to 14dB (a factor of 5) was successfully obtained in a wide frequency range from 600 GHz to 4 THz corresponding to the fundamental plasmon

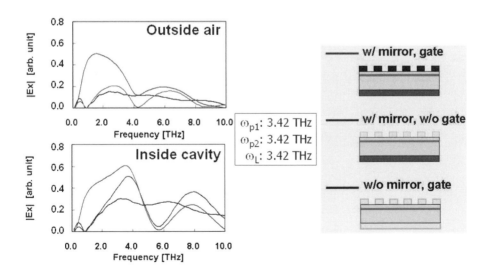

Fig. 6. Simulated frequency responses for three different device structures to impulsive excitation at all the plasmon cavities. Electric fields (x component) at two points (inside the cavity and outside air) are calculated by using a Maxwell's FDTD simulator and their temporal profiles were Fourier transformed. (after Ref. 6.)

resonance. One can see the fundamental peak stays at around 1.8 THz outside the air, which is fairly lower than the original characteristic frequency of ω_{p2}. One possibility for this cause would be the excitation of vertically coupled surface plasmon polaritons as is seen in the interfaces of metallic gratings[25]. It is noted that the wavelength under consideration is by two orders of magnitude larger than the feature size of the grating, which is thought to be a consequence of excitation of complex plasmon modes produced in the grating-bicoupled unique structure.

2.3. *Device fabrication*

The device was fabricated with InGaP/InGaAs/GaAs material systems in two structures: a standard single-heterostructure HEMT with metallic grating gates[7-11] and a double-decked (DD) HEMT with semiconducting two-dimensional electron gas (2DEG) grating gates[11, 23, 24]. The SEM (scanning electron microscopy) images for a typical metal-grating sample are shown in Fig.7(a). The 2D plasmon layer is formed with a quantum well at the heterointerface between a 15-nm thick undoped InGaAs channel layer and a 60-nm thick, Si-δ doped InGaP carrier-supplying layer. The grating gate was formed with 65-nm thick Ti/Au/Ti by a standard lift-off process. To cover operating frequencies from 1 to 10 THz, the grating geometry was designed with 350-nm G1 fingers and 100-nm G2 fingers to be aligned alternately with a space of 70 nm. The gate width is 75 μm for both G1 and G2. For comparison, another sample having a larger fraction in G1/G2 fingers (1800 nm/100 nm) was also fabricated. The number of gate fingers G1/G2 is 61/60 (38/37) for the sample having 300-nm (1800-nm) G1 fingers.

The device cross sectional view and its SEM image of a semiconducting grating-gate device are shown in Fig. 7(b)[23]. In this work, in order to produce the periodically-localized 2DEG, the double-decked HEMT structure is employed. The upper deck channel serves as the grating-gate antenna and is then periodically etched. Therefore more intensity in the emitted THz wave is expected. The HEMT structure consists of the InGaP/InGaAs/GaAs heterostructure with a selective doping in the InGaP layer. For the source/drain ohmic contacts, AuGe/Ni was lifted off and annealed after the upper-deck HEMT was selectively etched. The intrinsic device area is 30×75 μm^2, where the grating pattern is replicated on the upper-deck HEMT layer. The grating consists of 80-nm lines and 350-nm lines aligned alternately with a spacing of 100 nm. The number of fingers is 60 (61) for the 80-nm (350-nm) grating.

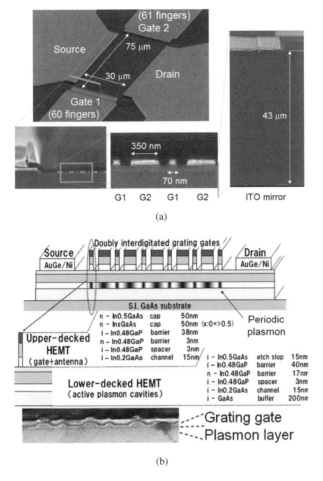

Fig. 7. SEM images of a fabricated metal grating-gate (a) and semiconducting grating-gate (b) plasmon-resonant emitter. (After Ref. 24.)

2.4. *Experimental Results and Discussions*

2.4.1. *DC-current-driven self oscillation*

Fourier-transformed far-infrared spectroscopic (FTIR) measurements were carried out for those samples[10, 11, 23, 24]. The samples were placed in the source position of the vacuum cavity of the FTIR. The radiation intensity was measured by Si bolometer having a responsivity of 2.84×10^5 V/W and a noise-equivalent power (NEP) 1.16×10^{-13} W/Hz$^{1/2}$. The experimental procedure was following – first we measured the background spectra – the spectra without any current flowing through the sample. This spectra contained information of the 300K blackbody emission modified by the spectral functions of all the elements inside the spectrometer. Then we measured the spectra with current flowing through the sample, and then, normalized them to the background data.

FTIR measured emission spectra for metal-grating gate samples having L_{G1}/L_{G2} = 70 nm/1850 nm and L_{G1}/L_{G2} = 70 nm/350 nm are shown in Fig. 8 (a) and (b), respectively[10, 11, 24]. One can see relatively broad spectra starting from about 0.5 THz with maxima around 2.5 THz for the first sample (S1) and around 3.0 THz for the second one (S2); the grating geometry reflects the spectral profile. For both samples the emission dies off abruptly around 6.5 THz, which is thought to be due to the Reststrahlen band of optical phonon modes of the GaAs-based materials[10].

The emission intensity versus V_{DS} is shown in Fig. 8(c). One can see that the emission intensity has a threshold property against V_{DS} and has a super-linear (nearly quadratic) dependence on V_{DS}. It is considered that the former property reflects on the coherent plasmons excited by the plasmon instability[3, 14, 15], while the latter property is attributed to the emission caused by the thermally excited incoherent plasmons due to injection of drifting hot electrons into the plasmon cavities[10, 24, 26, 27].

FTIR measured spectra for semiconducting-grating gate samples having L_{G1}/L_{G2} = 150 nm/1850 nm are shown in Fig. 9[11, 24]. It is, an intense emission power of the order of 1 μW for the DD-HEMT's at 300K (one order of magnitude higher than that for MGG-HEMT's). Dyakonov-Shur plasmon instability[3] and/or the Ryzhii-Satou-Shur transit-time instability[14, 15] may take place at the cavity boundaries where the electron drift velocity (thus, 2DEG density) modulation predominantly occurs. Analytical calculation suggests that the instabilities are critically promoted near the drain side with low 2DEG densities when V_{DS} exceeds the pinch off, resulting in the above-mentioned threshold property and enhancement of the emission at low frequency region around 2 THz . Since usually these coherent plasmons excitations are believed to have sharp spectral features, the observed spectral peaks may be attributed to these instability-driven emission.

The emission spectrum of thermally excited plasmons of the metallic grating-gate structure was calculated based on a first-principles electromagnetic approach elaborated upon in Refs. 26 and 27 where only a single metal grating in the structure is assumed as shown with solid lines in Fig. 8(a)[10]. In the calculated single-period structure spectrum, the fundamental plasmon resonance appears at around 3 THz and the second one at about 4.5 THz. This may explain a pronounced bump around 5 THz in the high-frequency shoulder in the short period structure experimental spectrum. The plasmon spectra in the actual double-grating structure in question, of course, must be more complex compared to that in a single-grating structure modeled numerically. In principle two different sorts of plasmon cavities can be formed under the metal fingers of different width in the double-grating structure.

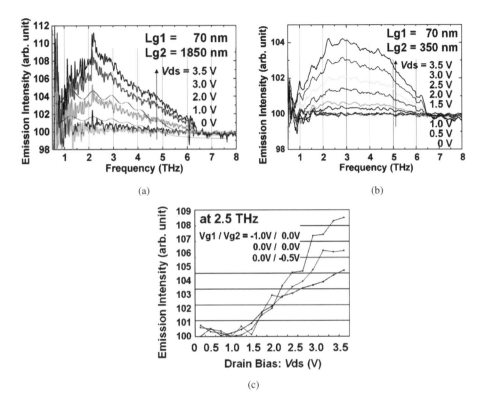

(a)   (b)

(c)

Fig. 8. FTIR measured emission spectra for metal-grating gate samples at room temperature. (a) L_{G1}/L_{G2} = 70 nm/1850 nm, (b) L_{G1}/L_{G2} = 70 nm/350 nm. The grating geometry reflects the emission spectra. (c) emission intensity at 2.5 THz vs. Vds for the sample with L_{G1}/L_{G2} = 70 nm/1850 nm, showing threshold behavior and super-linear (near quadratic) dependence on Vds. Two solid lines in (a) are calculated results for emissions from thermally excited plasmons at electronic temperatures of 310K and 320K. (after Ref. 24.)

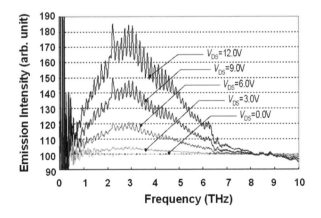

Fig. 9. FTIR measured emission spectra for a semiconducting grating-gate sample with L_{G1}/L_{G2} = 150 nm/1850 nm at room temperature. (after Ref. 24.)

2.4.2. *CW-pumped optically excited stimulated terahertz emission*

Next, the device was irradiated from the backside with a lineally-polarized 1550-nm band CW laser beam to measure its photoresponse at room temperature[8, 9]. A 1550-nm band tunable laser source with an average power of 2 mW was used. The polarization is set to be in parallel to the channel direction. Actually, the photon energy of the irradiated laser is much lower than all the band gap energies of this material system. However, the electrons are weakly photoexcited at the InGaAs/GaAs heterointerface via multi-step processes due to the existence of deep trap centers[27]. The photoelectrons are injected to the channel immediately due to the strong gate-to-channel electric field.

When the plasma wave resonance is excited, the DC drain-source potential is modulated because of the non-linear properties of the plasma fluid[4]. Therefore, the resonant intensity was measured by monitoring the DC modulation component ΔV_{DS} of the drain potential, which is called hereafter the photoresponse. The variation in ΔV_{DS} under irradiation was precisely lock-in amplified and detected at a chopping frequency of 1.29 KHz. At the same time, the terahertz radiation was detected using a 4.2-K cooled Silicon bolometer with a filter pass band from 0.6 to 3.5 THz.

Typical results for the V_{G1} and V_{G2} dependence of the photoresponses of a metal grating-gate sample (L_{G1}/L_{G2} = 300 nm/100 nm) for different V_{DS} conditions are shown in Fig. 10[8, 9]. The device exhibited a marked photoresponse with relatively sharp peaks on its V_{G2} dependence and with broad peaks on its V_{G1} dependence. For V_{G2} dependence, when V_{DS} = 1.0 V, the photoresponse exhibited a clear single peak at V_{G2} = -1.9 V. According to the Dyakonov-Shur model[4] this photoresponse peak at the lowest gate bias is interpreted as the fundamental plasmon resonance. With increasing in V_{DS} up to 3.0 V, the single peak becomes steeper and shifts up to higher V_{G2} point at around 0 V while the secondary peak grows up at lower V_{G2} point at -3.0 V, corresponding to the third-harmonic plasmon resonance. Similarly, for V_{G1} dependence, when V_{DS} increases from 1.0 to 3.0 V, weak double dips on the background slope at V_{DS} = 1.0 V grows up to clear double peaks at V_{G1} = -0.8 V and -2.8 V.

The results are completely different from that for standard HEMT devices having a single-gate finger fabricated on the same wafer showing monotonic dependence on the gate bias (plotted with a broken line in Fig. 10(a). Theoretical investigations in Refs. 3 and 13 suggest that the plasmon instability is promoted when electrons have a very high drift velocity of around 4×10^7 cm/s (equivalent plasma-wave Mach number of around 0.5). Carrier dynamics under weakly photoexcited conditions in our device are simulated based on an extended drift-diffusion model[29]. The result indicates that the injection of photoelectrons from a weakly charged 2DEG to the adjacent deeply charged 2DEG (plasmon cavity) is performed in a quasi-ballistic manner so that the above-mentioned instability condition is obtainable.

Focused on a simplest case at V_{DS} = 1.0 V, terahertz emission from the device was measured by using a 4-K cooled Si bolometer. At the same time, the device photoresponse was also measured. The measured results are plotted onto the V_{G1}-V_{G2} space as shown in Fig. 11(a) and (b)[9]. The photoresponse in Fig. 11(a) exhibits local

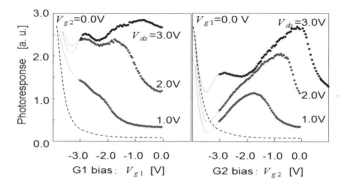

Fig. 10. Measured photoresponse to a single CW laser irradiation for a metal grating-gate sample having a grating gates geometry of L_{G1}/L_{G2} = 300 nm/100 nm. Dotted line in the left plot is for a standard HEMT, showing monotonic decrease with increase in gate bias. (after Ref. 9.)

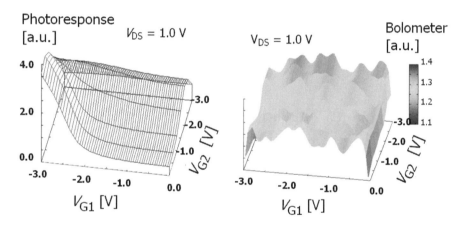

Fig. 11. (a) photoresponse and (b) 4K-cooled Si bolometer signal at V_{DS} =1.0 V for a metal grating-gate sample having a grating gates geometry of L_{G1}/L_{G2} = 300 nm/100 nm. (after Ref. 9)

maxima at $V_{G2} \approx$ -2.0 V along with the V_{G1} axis. In such a region, the bolometer shows clear enhancement of the signal. The observed signal is low and noisy due to atmospheric vapor absorption between the sample and the bolometer. According to well recognized responsivity on the order of 10^5 to 10^6 V/W for the Si composite bolometer used in this experiment (not calibrated) and atmospheric vapor absorption along with 20-cm propagation from the device to the bolometer, the emission power is roughly estimated to be 0.1 μW from a single device. When V_{G1} = -3 V and V_{G2} = 0 V, on the contrary, the photoresponse shows an increase but the bolometer doesn't detect the radiation. In this case, the photoresponse shows a non-resonant detection near to the threshold voltage. This phenomenon is well known as a space-charge effect of photoconductivity[30, 31]. Analytical calculation[3, 4, 9] indicates that the emission at a frequency around 1.3 to 1.8 THz should occur when $V_{G2} \approx$ -2.5 V and V_{G1} = 0 V for this sample, supporting the measured results.

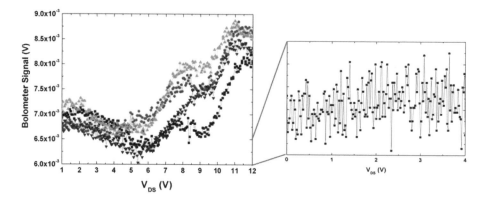

Fig. 12.   Left: bolometer detection of emission from a semiconducting grating-gate sample with L_{G1}/L_{G2} = 150 nm/1850 nm at room temperature as a function of V_{DS}. Measurement took place four times. Right: the result from a metal grating-gate sample with L_{G1}/L_{G2} = 75 nm/350 nm for comparison. (after Ref. 23.)

The result of the bolometric measurement for a semiconducting grating-gate sample is shown in Fig. 12 as a function of V_{DS}[11, 23]. The V_{DS} increases to the knee voltage from which the transistor is operated in the saturation region. The bolometer signal starts increasing at around 6 V and two clear peaks are observed at 8 and 11 V. These features were observed with good reproducibility as shown in Fig. 12. Compared with the results for metal grating-gate sample, remarkable enhancement in emission intensity by one order of magnitude was obtained. Note that the V_{DS} range is larger than that for metal grating-gate samples because the double-decked HEMTs in this work suffer from large parasitic source and drain resistance. Nevertheless the double-decked device exhibits more drastic change in the bolometer signal with increasing V_{DS}. This result supports the idea of low-conductive gate stack to enhance the THz radiation efficiency[6, 22], and therefore indicates that the proposed double-decked HEMT structure is a promising candidate to realize solid-state THz-wave emitters with high power and large efficiency.

It is inferred, from such a phenomenological coincidence that the marked photoresponse of this work is attributed to the plasmon excitation due to the injection of photoelectrons accelerated by the strong electric field arisen at the plasmon cavity boundaries, leading to self-oscillation of emission of terahertz electromagnetic radiation. Significant improvement on the plasmon resonance is owing to the original dual-grating gate device structure.

2.4.3.   *Two-photon injection-locked difference-frequency terahertz emission*

We conducted a so-called photomixing experiment. This is a trial for two-photon injection-locked difference-frequency terahertz generation. A pair of 1550-nm band tunable laser sources with an average power of 2 mW was used. The frequencies of those two laser sources are set so as to have a specific difference frequency Δf in the terahertz range. The polarizations of both beams are aligned be in parallel to the channel direction.

The illumination scheme is similar to that mentioned in 2.4.2. The emission spectra are measured using FTIR as described in 2.4.1.

The photogenerated electrons including terahertz Δf component are injected to the plasmon cavities. If Δf is close to the plasmon resonant frequency, the oscillation at Δf is promoted so that the neighboring frequency components may tend to be attracted to Δf. One can expect this gives rise to injection-locked coherent, monochromatic Δf generation.

The device used is a metallic grating gate type having a grating geometry of 70 nm and 1850 nm. First obtained preliminary results are shown in Fig. 13. Δf is set at 2.2 THz and 4 THz. For both cases, unfortunately, we could not yet succeed in injection-locking operation. However, when Δf is set at 2.2 THz which is close to 2.4 THz the background peak emission frequency, asymptotic behavior of injection locking to the Δf point is clearly observed. When Δf is detuned to 4 THz, the effect of frequency attraction is still observed but becomes weak.

The factors preventing from injection-locked oscillation are considered to be poor injection efficiency due to (i) broadband, intense background emission, and (ii) markedly long photoelectron life time (on the order of ns). Adding to the broadband hot plasmons emission dominating at higher V_{ds} as discussed in 2.4.1, dispersion of sheet carrier density along with the source-drain direction depending on applied V_{ds} is another factor that disperses the emission spectrum. Relatively low V_{ds} will help suppress this spectral broadening but the instability also weakens. Introduction of non-equal finger-size grating compensating for the electron density dispersion would be a solution. As mentioned in 2.4.2, the GaAs-based material systems used in the present device is transparent to the 1550-nm photons so that two-photon process via deep trap centers at the InGaAs/GaAs heterointerface takes place. Those photoelectrons have considerably long carrier life time which significantly attenuates the terahertz Δf photoelectron components. Introduction of InP material systems would be a solution.

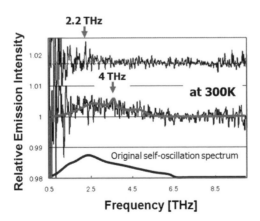

Fig. 13. Emission spectra when the device is subject to two-photon laser illumination at 300K, exhibiting asymptotic behavior of injection locking to the difference frequency (Δf) point. Δf is set at 2.2 THz (upper) and 4 THz (middle). The lower curve is an original self-oscillation spectrum without optical excitation (not scaled vertically).

2.4.4. *Impulsive laser excited terahertz emission*

Electromagnetic response to impulsive photoexcitation was also measured at room temperature by using reflective electrooptic sampling[7, 9]. A 1550-nm, 1-mW, 70-fs laser pulse was used as pump and probe beams. When the sample was appropriately biased, as shown in Fig. 14(a), it emitted an impulsive radiation followed by monochromatic relaxation oscillation which was significantly enhanced by its vertical cavity structure. The Fourier spectrum exhibited resonant peaks at 0.8 THz and its harmonic frequencies of up to 3.2 THz as shown in Fig. 14(b). These results are attributed to the emission of coherent electromagnetic radiation stimulated by photo-induced plasmon instability. Estimated radiation power would exceed 0.1 μW. We confirmed that the emission spectrum traces the 2D plasmonic dispersion relation in terms of its 2D electron density and the cavity size[8, 9].

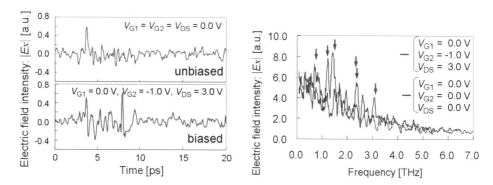

Fig. 14. Field emission response to impulsive photoexcitation. Grating gates geometry: L_{G1}/L_{G2} = 350 nm/70 nm. Left: temporal response, right: Fourier spectra. (after Ref. 9.)

3. Terahertz Intensity Modulator Based on Controlling 2D Plasmon Dispersion

One can consider the device structure as a gated functional element as shown in Fig. 15[32-35]. Suppose that the data signal is input to one gate grating so as to modulate the sheet electron density of the plasmonic cavity grating, and that a terahertz carrier wave is input to the device. If the intensity of the transmitted wave is sufficiently modulated by the gate bias, the data can be coded onto the terahertz carrier wave. Due to the high-speed nature of the HEMT structure, it is easy to perform the data coding at tens of Gbit/s onto the terahertz carrier[33]. Based on the above mentioned 2D-plasmonic nanostructure, we numerically analyze the dispersive effect on the transmission spectrum for coherent terahertz electromagnetic (EM) waves, and demonstrate functionalities of terahertz frequency multipliers as well as intensity modulators.

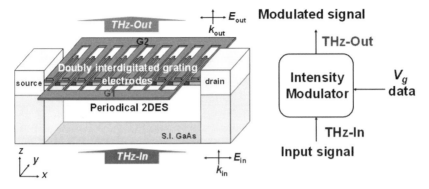

Fig. 15. Double-grating plasmonic nanostructure configuring intensity modulator.

Physical phenomena of 2D plasmons are described by the hydrodynamic equations. According to the Mikhailov's theory[13], after liberalized approximation the permittivity and conductivity of 2D plasmons are led to the following form[13, 33]:

$$\varepsilon(k,\omega) = \varepsilon_\infty + \frac{\pi dk^2 ne^2}{\varepsilon_0 m} \frac{i}{(\omega - kv_d)(\omega - kv_d + i\tau^{-1})}$$

$$\sigma(k,\omega) = \frac{ne^2}{m} \frac{i\omega}{(\omega - kv_d)(\omega - kv_d + i\tau^{-1})}$$

(3)

where ε_∞ the permittivity at finite frequency, ε_0 the permittivity in vacuum, d the gate-channel distance, n the density of electrons, e the electronic charge, m the effective mass, v_d the drift velocity, τ the total momentum relaxation time, ω the angular frequency, k the wave vector of the grating geometry. This Mikhailov's dispersive plasmonic conductivity is similar to the Drude-optical conductivity including n and ω, but is different and featured by two distinctive parameters having k and v_d. Based on Eq. (3), we can electronically control the dispersion by changing n and v_d due to the relation of the gate and drain bias voltages. We will implement the conductivity Eq. (3) into our in-house Maxwell's FDTD (Finite-Differential Time-Domain) simulator.

Figure 16 shows typical spatial field distributions of the electric field intensity (E_x) on the device cross section at a specific time step[32, 33]. The electron drift velocity v_d is fixed at 2×10^7 cm/s. When n is set at a relatively low value (2.2×10^9 cm^{-2}), the electric field intensity distributes monotonically so that a *radiative* mode of transverse-electric (TE) waves is excited. This is because, in this case, the fundamental mode of plasmons is dominantly excited to be coupled with the *zero* mode of TE waves. On the other hand, when n is set at a relatively high value (1.6×10^{12} cm^{-2}), anti-parallel electric field is excited with respect to the center of the channel. The electric filed intensity is cancelled out and *non-radiative* mode of TE waves is developed. This is because, in this case, the second harmonic mode of plasmons is predominantly excited to be coupled with the *first* mode of the TE waves. These results imply that the *radiative* or *non-radiative* mode is directly reflected by the density of electrons in the plasmon cavities. The mode coupling property is also affected by v_d. With increasing v_d, the threshold value of n at which the mode coupling shifts from *radiative* to *non-radiative* goes high. Detail discussion will be given in Ref. 35.

Figure 17 summarizes the field emission spectra for various n_{sp} conditions. In the frequency range from 0 to 4 THz, the field intensity decreases with increasing n_{sp}[33]. We speculate that transmission property for terahertz electromagnetic waves is markedly modulated by controlling n_{sp}, thus, the gate bias voltage. We further investigate the frequency dependence of the modulation efficiency or extinction ratio of the device. Figure 18 shows the typical results[34]. The field intensity is modulated by n_{sp}. With

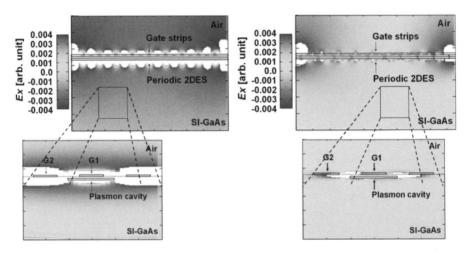

Fig. 16. Simulated spatial distribution of electric field intensity E_x at $n_{sp} = 2.2 \times 10^9$ cm^{-2} (left) and $n_{sp} = 1.6 \times 10^{12}$ cm^{-2} (right). $v_d = 2 \times 10^7$ cm/s. (after Ref. 33.)

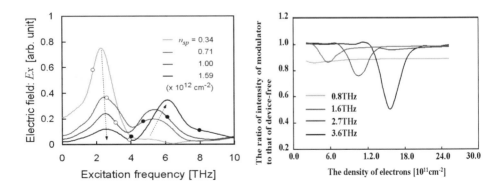

Fig.17. Field emission spectra for various n_{sp} conditions ○: fundamental plasmon mode, ●: 2nd harmonic plasmon mode. (after Ref. 32.)

Fig. 18. Normalized intensity of transmitted electromagnetic waves vs. n_{sp} for various terahertz carrier frequencies. (after Ref. 34.)

increasing the frequency of the EM wave, its intensity modulation can be performed in higher n_{sp} region with higher modulation efficiency. The extinction ratio defined as the ratio of the maximum to the minimum intensity is 3% at 0.8 THz, 15% at 1.6 THz, 24% at 2.7THz and 49% at 3.6THz, respectively. Note that the result for 0.8 THz showing a flat-band response with less intensity is due to the surface plasmon resonance at the grating gates.

Figure 17 also suggests another important aspect. For each spectrum, according to n_{sp}, the fundamental and 2nd harmonic plasmon resonances stay around 2~4 THz and 4~8 THz, respectively. Thus, it is seen that with increasing n_{sp}, the fundamental mode is suppressed while the 2nd mode becomes dominant. Therefore, the device can act as a terahertz frequency multiplier[32]. The above results demonstrate the electrically controllable dispersive effects on the transmission spectrum of a plasmon-resonant grating-gate HEMT device giving rise to potential functionality of a terahertz signal processing.

4. Conclusion

Recent advances in novel plasmonic nanotransistors for emission and intensity modulation of terahertz electromagnetic waves were reviewed. Material- and structure-dependent highly dispersive systems were configured with submicron-to-nanometer scaled 2D grating structures in a HEMT device. Analytical and Experimental studies revealed its various potential functionalities including emission, detection, intensity modulation, as well as frequency multiplication in the exploring terahertz frequency regime.

Acknowledgements

The authors would like to acknowledge Prof. T. Asano at Kyuhsu University, Japan, and all the students in Otsuji and Sano Laboratories for their extensive contributions throughout this work. They also thank Prof. W. Knap at Tohoku University, Japan (on leave from CNRS, France), Drs. D. Coquillat and F. Teppe at GES-UMR5650, CNRS and Montpellier 2 University, France for their experimental support and Prof. Victor Ryzhii at Univ. of Aizu, Japan and Prof. V.V. Popov and Dr. G.M. Tsymbalov at Inst. Radio Eng. Electron., Saratov Branch, Russia for their theoretical support, and Prof. K. Narahara at Yamagata University, Japan, Prof. M. Dyakonov at Montpellier 2 University, France, and Prof. M. Shur at Rensselaer Polytechnic Institute for many helpful discussions. This work was financially supported in part by the SCOPE Programme from the MIC, Japan, and by the Grant in Aid for Basic Research (S) from the JSPS, Japan.

References

1. M. Tonouchi, "Cutting-edge terahertz technology," Nature Photon., **1**, 97-105(2007).
2. H.-T. Chen, W.J. Padilla1, J. M.O. Zide, A.C. Gossard, A.J. Taylor and R.D. Averitt, "Active terahertz metamaterial devices," Nature, **444**, 597-600(2006).

3. M. Dyakonov, and M. Shur, "Shallow water analogy for a ballistic field effect transistor: new mechanism of plasma wave generation by dc current," Phys Rev Lett **71**, 2465-2468 (1993).

4. M. Dyakonov, and M. Shur, "Detection, mixing, and frequency multiplication of terahertz radiation by two dimensional electronic fluid," IEEE Trans. Electron Devices **43**, 380-387 (1996).

5. J. Lusakowski, "Nanometer transistors for emission and detection of THz radiation," Thin Solid Films **515**, 4327-4332 (2007).

6. T. Otsuji, M. Hanabe, T. Nishimura, and E. Sano, "A grating-bicoupled plasma-wave photomixer with resonant-cavity enhanced structure," Opt. Express **14**, 4815-4825 (2006).

7. T. Otsuji, Y. M. Meziani, M. Hanabe, T. Ishibashi, T. Uno, and E. Sano, "Grating-bicoupled plasmon-resonant terahertz emitter fabricated with GaAs-based heterostructure material systems," Appl. Phys. Lett. **89**, 263502 (2006).

8. Y. M. Meziani, T. Otsuji, M. Hanabe, T. Ishibashi, T. Uno, and E. Sano, "Room temperature generation of terahertz radiation from a grating-bicoupled plasmon-resonant emitter: Size effect ," Appl. Phys. Lett. **90**, 061105 (2007).

9. T. Otsuji, Y.M. Meziani, M. Hanabe, T. Nishimura, and E. Sano, "Emission of terahertz radiation from InGaP/InGaAs/GaAs grating-bicoupled plasmon-resonant emitter," Solid-State Electron. **51**, 1319-1327 (2007).

10. Y. M. Meziani, H. Handa, W. Knap, T. Otsuji, E. Sano, V. V. Popov, G. M. Tsymbalov, D. Coquillat, and F. Teppe, "Room temperature terahertz emission from grating coupled two-dimensional plasmons," Appl. Phys. Lett. **92**, 201108 (2008).

11. T. Otsuji, Y. M. Meziani, T. Nishimura, T. Suemitsu, W. Knap, E. Sano, T. Asano, V.V. Popov, "Emission of terahertz radiation from dual-grating-gates plasmon-resonant emitters fabricated with InGaP/InGaAs/GaAs material systems," J. Phys. Cond. Matt. **20**, 384206 (2008).

12. R.J. Wilkinson, C.D. Ager, T. Duffield, H.P. Hughes, D.G. Hasko, H. Armed, J.E.F. Frost, D.C. Peacock, D.A. Ritchie, A.C. Jones, C.R. Whitehouse, and N. Apsley, "Plasmon excitation and self-coupling in a bi-periodically modulated two-dimensional electron gas," J. Appl. Phys., **71**, 6049-6061(1992).

13. S.A. Mikhailov, "Plasma instability and amplification of electromagnetic waves in low-dimensional electron systems ," Phys. Rev. B, 58, 1517-1532 (1998).

14. V. Ryzhii, A. Satou, and M. Shur, "Plasma instability and amplification of electromagnetic waves in low-dimensional electron systems ," IEICE Trans. Electron. **E89-C**, 1012-1019 (2006).

15. V. Ryzhii, A. Satou, M. Ryzhii, T. Otsuji, and M. S. Shur, "Mechanism of self-excitation of terahertz plasma oscillations in periodically double-gated electron channels," J. Phys. Cond. Matt. **20**, 384207 (2008).

16. R. A. Hopfel, E. Vass, and E. Gornik, "Thermal excitation of two-dimensional plasma oscillations," Phys. Rev. Lett. **49**, 1667 (1982).

17. D. C. Tsui, E. Gornik and R. A. Logan, "Far infrared emission from plasma oscillations of Si inversion layers ," Solid State Comm. **35**, 875-877 (1980).

18. N. Okisu, Y. Sambe, and T. Kobayashi, "Far-infrared emission from two-dimensional plasmons in AlGaAs/GaAs heterointerfaces," Appl. Phys. Lett. **48**, 776-778 (1986).

19. R. Hopfel, G. Lindemann, E. Gornik, G. Stangl, A. C. Gossard and W. Wiegmann, "Cyclotron and plasmon emission from two-dimensional electrons in GaAs," Surf. Sci. **113**, 118-123 (1982).

20. K. Hirakawa, K. Yamanaka, M. Grayson and D. C. Tsui, "Far-infrared emission spectroscopy of hot two-dimensional plasmons in Al0.3Ga0.7As/GaAs heterojunctions," Appl. Phys. Lett. **67**, 2326-2328 (1995).
21. M. Hanabe, N. Imamura, T. Uno, T. Ishibashi, Y. M. Meziani and T. Otsuji, "Effects of parasitic capacitance on the terahertz plasmon resonance in GaAs MESFET's," in Extended Abstracts of the International Workshop on Terahertz Technology 2005, 181-182 (2005).
22. M. Hanabe, T. Nishimura, M. Miyamoto, T. Otsuji and E. Sano, "Structure-sensitive design for wider tunable operation of terahertz plasmon-resonant photomixer," IEICE Trans. Electron. **E89-C**, 985-992 (2006).
23. T. Suemitsu, Y.M. Meziani, Y. Hosono, M. Hanabe, T. Otsuji, E. Sano, "Novel plasmon-resonant terahertz-wave emitter using a double-decked HEMT structure" 65th Device Research Conference (DRC) Dig., 157-158 (2007).
24. T. Nishimura, H. Handa, H. Tsuda, T. Suemitsu, Y.M. Meziani, W. Knap, T. Otsuji, E. Sano, V. Ryzhii, A. Satou, V.V. Popov, D. Coquillat, and F. Teppe, "Broadband Terahertz Emission from Dual-Grating Gate HEMT's -Mechanism and Emission Spectral Profile," 66th Device Research Conference (DRC) Dig. 236-237 (2008).
25. J. A. Porto, F. J. Garchia-Vidal, and J. B. Pendry, "Transmission resonances on metallic gratings with very narrow slits," Phys. Rev. Lett. **83**, 2845-2848 (1999).
26. V. V. Popov, O. V. Polischuk, T. V. Teperik, X. G. Peralta, S. J. Allen, N. J. M. Horing, and M. C. Wanke, "Absorption of terahertz radiation by plasmon modes in a grid-gated double-quantum-well field-effect transistor," J. Appl. Phys. **94**, 3556-3562 (2003).
27. V. V. Popov, G. M. Tsymbalov, and N. J. M. Horing, "Anticrossing of plasmon resonances and giant enhancement of interlayer terahertz electric field in an asymmetric bilayer of two-dimensional electron strips," J. Appl. Phys. **99**, 124303 (2006).
28. T. Otsuji, M. Hanabe and O. Ogawara, "Terahertz plasma wave resonance of two-dimensional electrons in InGaP/InGaAs/GaAs high-electron-mobility transistors," Appl. Phys. Lett. **85**, 2119 (2004).
29. E. Sano, "Simulation of carrier transport across heterojunctions based on drift–diffusion model incorporating an effective potential," Jpn. J. Appl. Phys. **41**, L1306–L1308 (2002).
30. Y. Takanashi, K. Takahata, Y. Muramoto, "Characteristics of InAlAs/InGaAs high-electron-mobility transistors under illumination with modulated light," IEEE Trans. Electron Dev. **46**, 2271-2277 (1999).
31. M.S. Shur, J.-Q. Lu, "Terahertz sources and detectors using two dimensional electronic fluid in high electron-mobility transistors," IEEE Trans. Microwave Theory Tech. **48**, 750-756 (2000).
32. T. Nishimura, M. Hanabe, M. Miyamoto, T. Otsuji, and E. Sano, "Terahertz plasma wave resonance of two-dimensional electrons in InGaP/InGaAs/GaAs high-electron-mobility transistors," IEICE Trans. Electron. **E89-C**, 1005-1011 (2006).
33. T. Nishimura and T. Otsuji, "Terahertz polarization controller based on electronic dispersion control of 2D plasmons," International Journal of High Speed Electronics and Systems **13**, 547-555 (2007).
34. T. Nishimura, K. Horiike, and T. Otsuji, "A novel intensity modulator for terahertz electromagnetic waves utilizing plasmon resonance in grating-gate HEMT's," 7th Topical Workshop on Heterostructure Microelectron., 67-68 (2007).
35. T. Nishimura, N. Magome, and T. Otsuji, "A novel intensity modulator for terahertz electromagnetic waves utilizing two-dimensional plasmon resonance in a dual-grating-gate high electron mobility transistor," submitted to Jpn. J. Appl. Phys.

International Journal of High Speed Electronics and Systems
Vol. 19, No. 1 (2009) 55–67
© World Scientific Publishing Company

MILLIMETER WAVE TO TERAHERTZ IN CMOS

K. K. O, S. SANKARAN, C. CAO, E.-Y. SEOK, D. SHIM, C. MAO and R. HAN

Silicon Microwave Integrated Circuits and Systems Research Group, Department of Electrical and Computer Engineering, University of Florida, 539 New Engineering Building, Gainesville, Florida 32611, USA
kko@tec.ufl.edu

The feasibility of CMOS circuits operating at frequencies in the upper millimeter wave and low sub-millimeter frequency regions has been demonstrated. A 140-GHz fundamental mode VCO in 90-nm CMOS, a 410-GHz push-push VCO in 45-nm CMOS, and a 180-GHz detector circuit in 130-nm CMOS have been demonstrated. With the continued scaling of MOS transistors, 1-THz CMOS circuits will be possible. Though these results are significant, output power of signal generators must be increased and acceptable noise performance of detectors must be achieved in order to demonstrate the applicability of CMOS for implementing practical terahertz systems.

Keywords: CMOS; mm-wave; oscillator; Schottky diode; detector; phase locked loop; terahertz

1. Introduction

Electro-magnetic waves in the frequency range between 100 and 600 GHz have been utilized in spectroscopy, in active and passive imaging for detection of concealed weapons, chemicals and biological agents, and in short range radars and secured high data rate communications[1-3]. Typically compound semiconductor devices are employed to construct such systems, and the high cost and low level of integration of those devices have limited the growth of these applications. Recent progress in CMOS (Complementary Metal Oxide Silicon) integrated circuit (IC) technology has made it possible to consider CMOS as an alternative means for realization of capable and economical systems that operate at 200 GHz and higher. This paper discusses the performance of devices available in digital CMOS as well as those of signal sources and detectors operating between 100 and 410 GHz fabricated in digital CMOS which suggest the feasibility of terahertz operation of CMOS circuits.

2. Transistors and Diodes in CMOS

2.1. *Speed Performance of NMOS Transistor*

The consideration for possible use of CMOS to fabricate terahertz systems is enabled by the technology scaling of CMOS. Figure 1 shows the projected performance requirements of NMOS transistors and InP hetero junction bipolar (HBT's) transistors in manufacturing. These plots are extracted from the 2006 International Roadmap for

Semiconductors (ITRS) [4]. By year 2013, the projected NMOS unity power gain frequency (f_{max}) requirement is ~650 GHz. With such transistors, it will be possible to build an amplifier operating up to 300 – 350 GHz. Another interesting point is that by 2011, the projected f_{max} requirement of NMOS transistors in production will be higher than that of InP HBT's in production. This change is not driven by the change in electronic properties of silicon nor InP. This change is driven by the economy of scale and demand for CMOS technology that are enabling scaling of NMOS transistors to the nano-meter regime.

There is great deal of concerns whether this scaling of CMOS technology can continue. Figure 1 also shows the measured f_T and f_{max} of transistors from process technologies which are believed to be in production. A bulk transistor from a 65-nm technology[5] has f_{max} of 420 GHz, while an SOI transistor from a 45-nm process has f_T of 450 GHz[6]. These suggest that the industry has been able to keep up with the ITRS to date. Presently, it is not clear whether the industry will be able to keep up with ITRS to 2020. Nevertheless, most likely, the situation will not drastically change in the near term and this will enable amplification at 300 GHz higher using NMOS transistors within next two to three years.

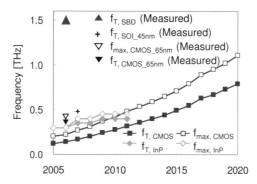

Figure 1. The projected performance requirements of NMOS transistors from 2006 International Road Map for Semiconductors[4] and the data from the literature.

2.2. *Schottky Diode and Other devices for THz detection*

In the near term, at frequencies higher than ~400 GHz, it will not be possible to achieve amplification using NMOS transistors. A way to deal with this is to use passive detectors and frequency multipliers as presently done routinely for sub-millimeter and THz electronic systems[2] based on III-V devices. In particular, Schottky diodes are widely used for this purpose. It turns out that it is possible to implement THz diodes in CMOS without any process modifications. Figure 2 shows the layout of one cell and a cross section. The Schottky contact is formed on a diffusion region where there are no source/drain implantations. The ohmic contacts placed on the n^+ implanted n-well form the n-terminal. The metal connections to the Schottky and ohmic contacts are spaced wide apart to reduce the parasitic capacitance. Such a diode with $CoSi_2$-silicon Schottky junction has been realized in a 130-nm CMOS process[7]. The Schottky contact area is set at the minimum to maximize the cutoff frequency. This type of diodes can also be used for frequency multiplication to generate sub-millimeter wave signals[2,8].

The detector diode was formed using sixteen 0.32μm x 0.32μm cells connected in parallel (R_s=13 Ω and C=8 fF). Figure 3 shows the cut-off frequency versus the diode bias voltage. At zero bias, the cut-off frequency is ~1.5 THz for the $CoSi_2$ to n-well diode. With further optimization, it may be possible to increase the cut-off frequency to ~2 THz. With such diodes, it should be possible to build detectors operating above 500 GHz.

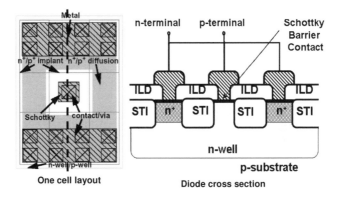

Figure 2. Integrated Schottky diode.

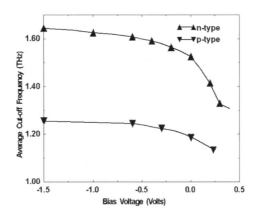

Figure 3. Cut-off frequencies of $CoSi_2$ to n-well and $CoSi_2$ to p-well diodes estimated from R and C measured around 20 GHz.

It is also possible to pick up terahertz signals using a plasmas wave detector fabricated with NMOS transistors. Such a detector using transistors with a gate length of 120-300 nm and gate oxide thickness of 1.1 nm has been demonstrated in CMOS. The minimum noise equivalent power of detector is 1×10^{-10} W-Hz$^{-0.5}$ at 700 GHz[9]. The result suggests that there maybe an alternate way to detect terahertz waves using a conventional device available in CMOS.

3. Signal Sources

A fundamental mode voltage controlled oscillator (VCO) that achieves 140-GHz operation[10] and a push-push VCO that achieves 410-GHz operation[11] were demonstrated. A critical issue for a signal source is stability. It must be locked using a phase locked loop (PLL). A 50-GHz charge pump PLL utilizing an LC oscillator based injection locked divider that provides a second order harmonic output at 100 GHz was demonstrated[12]. This was the first CMOS PLL operating in the millimeter wave band.

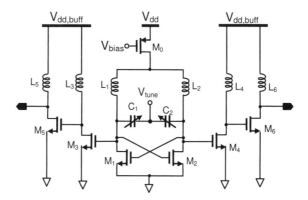

Figure 4. Schematic of 140-GHz voltage controlled oscillator.

3.1. *A. Fundamental Mode VCO*

Figure 4 shows the schematic of 140-GHz VCO fabricated in a 90-nm logic CMOS process[10]. It employs LC-resonators. Cross-coupled NMOS transistors (M_1, M_2) form the VCO core. Inductors (L_1, L_2), accumulation mode MOS capacitors/varactors (C_1, C_2), and the capacitances associated with M_1 and M_2 form the LC resonators. The width of cross-coupled transistors is 8.36 μm. The varactors are formed by two 0.5 μm x 0.18 μm fingers with contacts on both sides of the gate finger. The inductors, L_1 and L_2, are built using a single loop circular inductor. The trace was formed using the top copper layer with a thickness of 0.8 μm. The trace width is chosen to be 2 μm. The diameter of circular inductor is 31.6 μm and the inductance of the loop is around 65 pH. The VCO utilizes lumped elements which are more compact than those based on transmission lines[13]. The maximum length of inductors is around 80 μm, which is less than 10% of the wavelength in SiO$_2$ at 140 GHz.

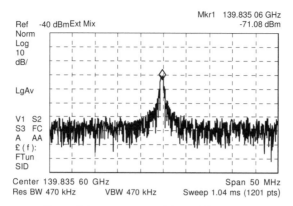

Figure 5. Output spectrum of the 140-GHz VCO.

Figure 5 shows an output spectrum. The conversion loss of harmonic mixer used for the measurements is about 50 dB and the insertion loss of probe is about 2 dB at 140 GHz. Thus, the output is estimated to be ~ -19 dBm. The measured phase noise is ~ -85 dBc/Hz at 2-MHz offset from the carrier. The output frequency can be tuned from 139 to 140.2 GHz by changing the varactor voltage and bias current[13]. The VCO draws 8 mA from a 1.2-V supply. Two VCO's with larger inductors are also included in the same chip. Their maximum operating frequencies are 110 and 123 GHz, and tuning ranges are 2.4 and 1.6 GHz, while the output power levels are -10 and -14 dBm, respectively. The tuning range is limited because the design is targeted for high operating frequency. The circuit can be re-optimized to achieve a wider tuning range at lower operation frequencies[13].

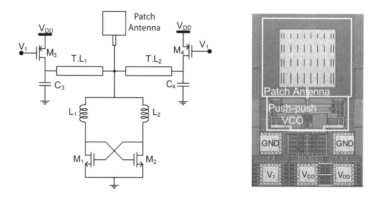

Figure 6. Schematic and photograph of a 410-GHz push-push VCO in 45-nm CMOS.

3.2. *Push-Push VCO*

Push-push VCO's [11, 14-19] are based on fundamental mode oscillators similar to the 140-GHz VCO. Figure 6 shows the schematic of 410-GHz push-push VCO[11] fabricated in 45-nm CMOS. At the virtual ground nodes, the anti-phase fundamental signals cancel out and the second harmonic signal can be extracted. The middle point of inductors L_1 and L_2 has the parasitic capacitance to ground with the lowest value and highest Q among the common-mode nodes. This makes the impedance at resonant frequency the highest and the best port to extract the push-push output[18]. The inductors L_1 and L_2 are implemented using a single 40pH differential inductor. To reduce the parasitic capacitance to substrate, only the top metal layer is used. The top metal layer is about 2 μm thick and about 2 μm above the silicon substrate. The diameter is about 20 μm and metal width is 1.6 μm. The polysilicon layer is used to form a patterned ground shield. A quarter wavelength transmission line tuned for the second harmonic frequency is usually used to increase the amplitude of second harmonic while suppressing the fundamental signal [14,18]. In this design, the transmission line and current source transistor are broken into two parts to make the layout symmetric to better suppress the fundamental signal at the common mode nodes.

The transmission line is formed using the grounded coplanar waveguide (GCPW) structure shown in Figure 7. Compared to the conventional CPW, the ground plane isolates the line from the lossy silicon substrate and reduces the insertion loss. The lines are formed using the top bond pad metal layer and the ground plane is formed by shunting metal layers 1-5. The chip occupies 640 μm x 390 μm including the bond pads. The die photograph is also shown in Figure 6.

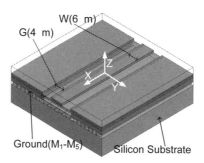

Figure 7. Grounded coplanar wave guide.

Measurements of ~400-GHz oscillator outputs are challenging. The output power of push-push oscillator decreases, and losses of interconnects and harmonics mixers increase with frequency. Presently, there are no commercially available probes for measurements at this frequency. Because of this optical techniques are utilized. To enable this, an on-chip patch antenna with an area of 200 x 200 μm² is integrated with the oscillator. The patch is formed using the bond pad metal layer while the ground plane is formed with metal layers 1-5 shunted together like the way GCPW is constructed. The patch and ground plane are separated by ~4 μm thick SiO₂ layer. The relatively thin gap reduces the efficiency of antenna to ~ 20%. Increasing the gap to 7 μm should increase the efficiency to ~50%.

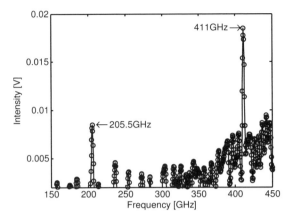

Figure 8. Output spectrum of the 410-GHz push-push VCO.

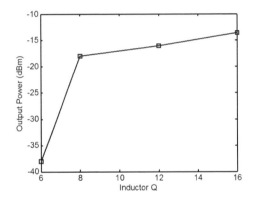

Figure 9. Simulated output power is sensitive to the Q of the inductor. When the Q at 205 GHz is increased to above 8, the output power can be increased to higher than -20 dBm.

The spectrum of VCO is measured using a Bruker 113V Fourier Transform Infrared Spectroscopy (FTIR) system. The power level was measured using a silicon bolometer (HD-3, 1378) from Infrared Laboratories. Figure 8 shows an output spectrum. The power level was measured using a bolometer. The measured signal level is -49 dBm for the signal at 411 GHz. The circuit consumes 17 mW. The oscillation frequency can be tuned from 409 to 413 GHz. The phase noise could not be measured using an FTIR. The power level is low. This is mostly due to the losses associated with the transistor and thin metal layers available in the CMOS process. Figure 9 shows the simulated oscillator output power versus inductor Q at 205 GHz. When the Q is increased to above 8, the output power can be increased to higher than -20 dBm. The inductor Q can be improved by increasing the thickness of top metal layer to 4 μm as well as dielectric layer between the top metal layer and silicon substrate to ~ 6 μm. This should also decrease the loss of transmission lines which should also increase the output power. This range of thicknesses is compatible with CMOS, and is already available in some silicon integrated circuits technologies optimized for mm-wave applications. The patch antenna and oscillator are originally designed for 390 GHz instead of 410 GHz. This mismatch between the tuned frequencies of antenna and oscillator reduces the output power by 2 dB. The output frequency of 410 GHz is the highest frequency for signals generated using transistors in any technologies including those based on compound semiconductors.

3.3. *Phase Locked Loop*

Generation of free running high frequency signal by itself is not sufficient. The signal must be stabilized using a phased locked loop (PLL). To address this, a fully integrated PLL tunable from 45.9 to 50.5 GHz, which also outputs the second order harmonic at frequencies between 91.8 and 101 GHz, has been demonstrated in a 130-nm logic CMOS process[12]. It consumes less than one-tenth of the power of a recently reported SiGe HBT PLL using static frequency dividers, while achieving comparable phase noise performance[20]. Figure 10 shows a block diagram of the 50-GHz PLL.

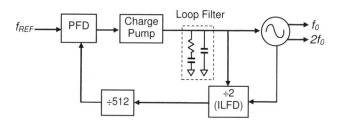

Figure 10. 50-GHz phase locked loop block diagram.

The circuit utilizes an injection-locked frequency divider (/2) (ILFD)[21], and a 1/512 static frequency divider. The PFD (Phase Frequency Detector) uses a three-state phase detection scheme. The LPF (Low pass filter) is 2nd order and the PLL is 3rd order. The loop filter consists of two MOS capacitors and one polysilicon resistor. The loop bandwidth target is ~500 kHz and the phase margin is ~70 degrees. The reference frequency for the PLL is ~50 MHz. The injection locked divider is an LC-tank type[22], which is essentially an oscillator running near the frequency a half of that for the VCO. Usually, the first frequency divider limits the maximum PLL operating frequency. Since, the divider is an oscillator whose operating frequency is lower than that of the VCO frequency, the PLL maximum operating frequency is limited by the maximum VCO frequency instead of that for the divider. Given the demonstration of a 192-GHz VCO in 130-nm CMOS and the fact that presently, the leading edge production technology is at 45-nm, PLL's operating around 400 GHz or higher should be possible in CMOS. Figure 11 shows the schematic of VCO and injection locked divider combination. The chip size is 1.16 x 0.75 mm^2 including bond pads.

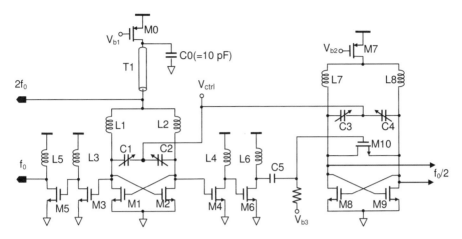

Figure 11. VCO and injection locked divider.

Table 1: Summary of measured phase locked loop characteristics

Locked Frequencies	45.9 – 50.5 GHz (fundamental) 91.8 – 101 GHz (second harmonic)
Phase Noise	-63.5 dBc/Hz @ 50 kHz offset -72 dBc/Hz @ 1 MHz offset -99 dBc/Hz @ 10 MHz offset
Output Power	-10 dBm (fundamental)
Settling Time	< 40 μs
Spurs	-40 to -27 dBc
Supply Voltage	1.5/0.8 V
Power Dissipation	57 mW
Die Area	1.16 x 0.75 mm²
Technology	UMC 0.13-μm logic CMOS

The measured performance of PLL is summarized in Table 1. The VCO starts to oscillate at 0.9-V supply voltage. To achieve higher output power and lower phase noise, the measurements are made at 1.5-V V_{DD} and 12-mA bias current. The output power level is about -10 dBm and the phase noise of free running VCO is about -90 and -109 dBc/Hz at 1-MHz and 10-MHz offset from the carrier, respectively. Figure 12 shows the locked spectrum of 2nd harmonic near 100 GHz. This is the signal with the highest frequency ever locked by a silicon circuit.

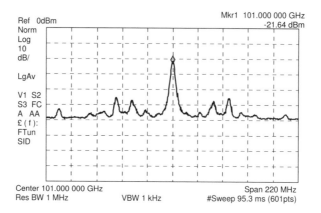

Figure 12. Phase locked second harmonic output ~100 GHz.

4. Detector Circuits

To implement a complete system, a detector is also required. To examine the feasibility of this, a 182-GHz Schottky diode detector was fabricated once again using 130-nm foundry CMOS[23]. To provide controlled input signals, the detector input was generated on-chip by modulating the bias current of a push-push VCO. As shown in Figure 13, the detector consists of a ~180-GHz RF matching circuit, a Schottky diode, a low pass filter, and an amplifier for driving a 50Ω load. The diode has been forward biased at 300 μA through a 1-kΩ resistor (R3). The detector input impedance is conjugate matched to the signal source impedance at ~180GHz. The low pass filter has a corner frequency of ~10GHz.

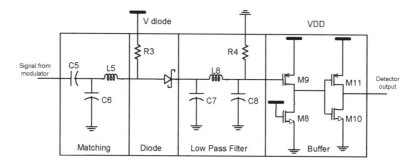

Figure 13. Detector schematic.

Figure 14 (left) shows a schematic of circuit for generating the modulated 182-GHz signal. This design utilizes the existing 192-GHz VCO[18]. Buffers were added to monitor the fundamental output. This lowered the frequency at push-push port to ~182 GHz. The amplitude of this VCO was modulated by changing the gate voltage of M3, or the bias current (I_{bias}) of the oscillator. The modulating signal is AC-coupled to M3 through a capacitor (C3). A shunt 50-Ω (R1) termination resistor is added to make the input amplitude at gate of M3 more predictable. The bypass capacitor C4 is an AC-short for the signal near 180GHz, but it presents high impedance to the modulating signal. Since changing I_{bias} also modifies the drain voltage of the VCO core transistors and thus the capacitances of the L-C tanks[18], the input signal also modulates the output frequency.

Figure 14 (right) shows the voltage waveforms of modulation and detected signals across a 50-Ω load, when the VCO is modulated with 10-MHz 0.1-V amplitude input signal at 1.75-V V_{DD}. The dc bias voltage at gate of M3 (V_{bias}) is 0V. The 10-MHz input frequency is chosen to attenuate the un-intended feed through of modulation signal via C5 (50fF). The detected signal frequency is the same as that for the modulation signal. Figure 15 is a photograph of the detector. The chip size is 1120 x 600 μm^2 including bond pads. This work suggests that Schottky diodes implemented in CMOS without any process modifications can be used to overcome the operating frequency limited by transistors.

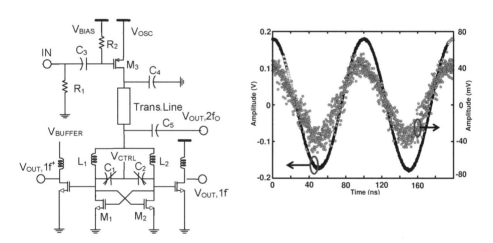

Figure 14. (left) Modulated signal generation circuit. (right) Measured 10-MHz modulation input signal compared to detector output signal.

5. Conclusions

The feasibility of using a mainstream foundry 130-nm logic CMOS process to fabricate a signal generator and a detector that operate in the vicinity of 200 GHz has been demonstrated. Using a 45-nm CMOS technology, 410-GHz signal has been generated, which is the highest ever using transistors of any technologies. The long term prospect is even more exciting. According to the 2006 International Roadmap for Semiconductor, by 2018, the required f_T and f_{max} of NMOS transistors in production are 0.7 and ~1 THz[4]. With such transistors and technology, it should be possible to implement circuits operating near 1THz. Attaining the terahertz operation frequency, though significant, is only a necessary step on the way to realize CMOS terahertz systems. To be sufficient, signal sources with output power greater than at least -20 dBm and detectors with room temperature noise equivalent power of 5×10^{-12}W/Hz$^{0.5}$ or less at frequencies greater than 300 GHz should be demonstrated. These are worthy research goals that can revolutionize how terahertz frequency regime is utilized.

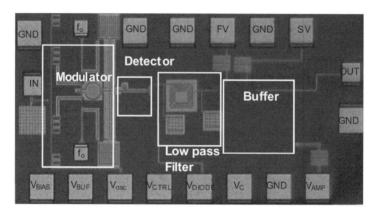

Figure 15. Photograph of 182-GHz detector circuit.

6. Acknowledgments

This work is supported by DARPA (N66001-03-1-8901) and SRC (Task 1327). The circuits are fabricated by UMC and TI.

References

1. P. H. Siegel, "THz technology," *IEEE Trans. on MTTS.*, vol. 50, no. 3, pp. 910-928,Mar. 2002.
2. T. W. Crow et al., "Opening the THz window with integrated diode circuits," *IEEE J. of Solid-State Circuits*, vol. 40, no. 10, pp. 2104-2110, Oct., 2005.
3. D. L. Woolard et al., "THz Freq. Sensing and Imaging, A Time of Reckoning Future Applications?" *IEEE Proc.*, vol. 93, no.10, pp. 1722 -1743, Oct. 2005.
4. *2006 International Roadmap for Semiconductors*, SIA, San Jose, CA.
5. I. Post et al., "A 65nm CMOS SOC Technology Featuring Strained Silicon Transistors for RF Applications," 2006 International Electron Device Meeting, Late News, San Francisco, Dec. 2006.
6. S. Lee et al., "Record RF Performance of 45-nm SOI CMOS Technology," 2007 International Electron Device Meeting, pp. 255-258, Washington D.C., Dec. 2007.
7. S. Sankaran, and K. K. O, "Schottky Barrier Diodes for mm-Wave and Detection in a Foundry CMOS Process," *IEEE Elec. Dev. Letts.*, vol. 26, no. 7, pp. 492-494, July 2005.
8. C. Mishra et al., "Silicon Schottky Diode Power Converters Beyond 100 GHz," 2007 *IEEE RFIC Symp.* pp. 547-550, June 2007.

9.   R. Tauk, F. Teppe, S. Boubanga, D. Coquillat, W. Knap, Y. M. Meziani, C. Gallon, F. Beouf, T. Skotnicki, D. K. Maude, S. Rumyantsev, and M. S. Shur, "Plasma wave detection of terahertz radiation by silicon field effect transistors: Responsivity and noise equivalent power," Applied Physics Letters, vol. 89, 23511 (2006).

10. C. Cao and K. K. O, "A 140-GHz Fund. Mode VCO in 90-nm CMOS Tech.," *IEEE Microwave and Wireless Comp. Letts*, vol. 16, no. 10, pp 555-557, Oct. 2006.

11. E. Y. Seok, C. Cao, D. Shim, D. J. Arenas, D. Tanner, C.-M. Hung, and K. K. O, "410-GHz CMOS Push-push Oscillator with a Patch Antenna," 2008 International Solid-State Circuits Conference, pp. 472-473, Feb. 2008, San Francisco, CA.

12. C. Cao, Y. Ding and K. K. O, "A 50-GHz PLL in 130-nm CMOS," 2006 *IEEE Custom Integrated Circuits Conference*., pp. 21-22, Sep. 2006, San Jose, CA.

13. C. Cao and K. K. O, "Millimeter-wave VCO's in 0.13-µm CMOS technology," *J. of Solid-State Circuits*, pp. 1297-1304, June, 2006.

14. Y. Baeyens and Y. Chen, "A monolithic integrated 150 GHz SiGe HBT push–push VCO with simultaneous differential V-band output," *MTT-S International Microwave Symposium Digest*, vol. 2, pp. 877–880, Jun., 8–13 2003, Philadelphia, PA.

15. R.C. Liu et al., "A 63GHz VCO using a standard 0.25µm CMOS process," *2004 International Solid State Circuits Conference*, pp. 446~447, Feb. 2004, San Francisco CA.

16. P.-C. Huang et al., "A 114GHz VCO in 0.13µm CMOS Technology," *2005 International Solid State Circuits Conference*, pp. 404~405, Feb. 2005, San Francisco, CA.

17. P.C. Huang et al., "A 131 GHz push-push VCO in 90-nm CMOS," *RFIC Symposium*, pp 613~616, June 2005, Long Beach, CA.

18. C. Cao, E. Seok and K. K. O, "192-GHz push-push VCO in 0.13-µm CMOS," *IEE Electronic Letts.*, vol. 42, no. 4, Feb. 2006, pp 208-209.

19. R. Wanner, R. Lachner, and G. R. Olbrich, "A monolithically integrated 190-GHz SiGe push–push oscillator," *Microwave and Wireless Components Letters*, vol. 15, no. 12, Dec. 2005, pp. 862-864.

20. W. Winkler et al., "A fully integrated BiCMOS PLL for 60 GHz wireless applications," *2005 International Solid State Circuits Conference*, pp. 406–407, Feb. 2005, San Fran., CA.

21. H.R. Rategh et al., "A CMOS freq. synthesizer with an injection-locked freq. divider for a 5-GHz WLAN receiver," *IEEE J. of Solid-State Circuits*, pp. 780–787, vol. 35, no. 5, May 2000.

22. K. Yamamoto and M. Fujishima, "55GHz CMOS freq. divider with 3.2GHz lock. range," *Proc. European Solid State Circuits Conference*, pp. 135–138, Oct. 2004.

23. E. Seok, S. Sankaran and K. K. O, "A mm-Wave Schottky Diode Detector in 130-nm CMOS," *Symposium on VLSI Circuits*, pp.178-179, June 2006, Honolulu, HI.

International Journal of High Speed Electronics and Systems
Vol. 19, No. 1 (2009) 69–76
© World Scientific Publishing Company

THE EFFECTS OF INCREASING AlN MOLE FRACTION ON THE PERFORMANCE OF AlGaN ACTIVE REGIONS CONTAINING NANOMETER SCALE COMPOSITIONALLY INHOMOGENEITIES

A.V. SAMPATH, M.L REED, C. MOE, G.A. GARRETT, E.D. READINGER, W.L SARNEY,
H. SHEN, M. WRABACK

US Army Research Laboratory, Adelphi MD 20783

C. CHUA, N.M. JOHNSON

Palo Alto Research Center, Palo Alto, CA 94304, USA

In this paper we report on the characterization of n-$Al_{0.51}Ga_{0.49}N$ active regions and the fabrication of ultraviolet LEDs that contain self-assembled, nanometer-scale compositional inhomogeneities (NCI-AlGaN) with emission at ~290 nm. These active regions exhibit reduced integrated photoluminescence intensity and PL lifetime relative to 320 nm NCI-AlGaN active regions that have significantly lower AlN mole fraction, despite having more than an order of magnitude fewer threading dislocations, as measured by transmission electron microscopy. This behavior is attributed to nonradiative recombination associated with the presence of a larger density of point defects in the higher Al content samples. The point defects are ameliorated somewhat by the lower density of NCI AlGaN regions in the higher Al content samples, which leads to a larger concentration of carriers in the NCI and concomitant reduced radiative lifetime that may account for the high observed peak IQE (~ 25%). Prototype flip chip double heterostructure-NCI- ultraviolet light emitting diodes operating at 292 nm have been fabricated that employ a 50% NCI-AlGaN active region.

1. Introduction

Semiconductor based ultraviolet light emitting devices have significant applications for air, water, and surface purification, as well as food storage and novel communication systems. III-Nitride semiconductors are attractive materials for this application owing to a direct band gap that is widely tunable from 6.1 eV (AlN) to 3.4 eV (GaN). Currently, III-Nitride based ultraviolet light emitting diodes (UVLEDs) are commercially available that operate at wavelengths as short as 247 nm. Nichia Inc. offers high power 365 nm UVLEDs that are capable of providing 250 mW output power at 500 mA DC current, based upon GaN multiple quantum well active regions[1]. Shorter wavelength UVLEDs emitting at 280 nm are offered by SET Inc. that are able to provide ~ 0.5 mW output power at 20 mA current[2]. This reduction in output power at shorter emission wavelengths is attributed to the need for III-Nitride alloys with increasing AlN mole fraction, which results in decreasing conductivity of the hole and electron injection layers and increasing non-radiative recombination due to increasing alloy fluctuations and defect density.

These factors combine to result in poorer wall plug efficiency and output power for UVLEDs at shorter wavelengths.

A strategy for improving the performance of UVLEDs is to suppress non-radiative recombination in the active region through the localization of carriers. Previously, we have reported on the development of AlGaN containing nanometer-scale compositional inhomogeneities (NCI-AlGaN) that exhibit intense room temperature photoluminescence at a peak wavelength that is significantly red-shifted with respect to the band gap of the alloy[3]. We attribute these phenomena to the concentration of carriers in nanometer-scale regions of narrower band gap alloy residing within a wider band gap matrix. This concentration of carriers saturates non-radiative states in these NCI-AlGaN films and promotes radiative recombination[4]. Using this approach we have fabricated double heterostructure UVLEDs (NCI-DH-UVLEDs) employing a NCI-AlGaN active region that emit at 321 nm and have an output power of 0.56 mW for a DC drive current of 90 mA, based upon wafer level testing[5]. In this paper, we report on the impact of increasing the AlN mole fraction in the NCI-AlGaN matrix on its performance as an UVLED active region by comparing heterostructures emitting at ~290 nm with those emitting at ~325 nm.

2. Method

The samples investigated consist of two heterostructures whose active regions emit at 323 nm and 292 nm. The double heterostructure emitting at 323 nm was grown on a 1 μm thick AlGaN film deposited by hydride vapor phase epitaxy (HVPE) on c-plane sapphire. The structure consists of a 50 nm thick NCI-AlGaN layer containing 33% AlN by mole fraction sandwiched between a 300 nm thick, Si- doped, n-AlGaN film having 40% AlN by mole fraction and a 6 nm thick, unintentionally n-type doped AlGaN film having 40% AlN by mole fraction, all deposited by plasma-assisted molecular beam epitaxy (PA-MBE). For simplicity this sample will be referred to herein as the 323 nm double heterostructure (323 nm-DH).

The 292 nm active region was deposited as a single heterostructure and was grown on a 2.3 μm thick, Si doped AlGaN template deposited by metalorganic chemical vapor deposition (MOCVD) having 55% AlN by mole fraction. The MBE grown structure consist of a 100 nm-thick Si doped AlGaN layer containing ~55% AlN by mole fraction followed by a 80 nm thick NCI-AlGaN active region layer containing 51% AlN by mole fraction. This sample will be referred to as the 292 nm single heterostructure (292 nm-SH).The growth process for depositing these structures on an AlGaN template has been reported elsewhere [6].

The optical properties of the samples were investigated by temperature dependent cw-photoluminescence using a 244 nm Coherent FReD Argon Ion Laser with a low output power of ~0.5 mW so as to avoid effects associated with the saturation of defects that can result from the excitation of a large density of photo-generated carriers. The samples were cooled to ~8K using a cryostat and a closed-cycle He compressor. The photoluminescence lifetimes of the heterostructures were measured by time-correlated single photon counting with ~25 ps resolution using a broadly tunable (230-375 nm) fs optical parametric amplifier as the excitation source. The structural properties of the samples were investigated by transmission electron microscopy.

3. Results

3.1. *Temperature dependent cw-Pl studies*

Figure 1 shows the temperature dependent cw- photoluminescence measurements for both of the heterostructures. The room temperature PL spectra for the 323 nm-DH (left) is dominated by an intense single peak emission at 323 nm and a pronounced low energy tail. Upon cooling the sample to 8K, the intensity of the PL spectral peak increases by about a factor of 4 and a high energy shoulder becomes more apparent that is attributed to both band edge emission from the active region as well as from the barrier layers in the double heterostructure. The internal quantum efficiency (IQE) of the active region can be estimated from this data as I_{293K}/I_{8K}, where I_{293K} and I_{8K} are the integrated PL intensities measured over the spectrum of the UV peak, assuming that non-radiative states are occupied and therefore frozen out at low temperature. This assumption was verified by demonstrating that the PL intensity from the NCI regions did not increase over the temperature range between 8 and 25K. Using this methodology, the IQE for the 323-DH is estimated to be ~22.4%, which is significantly higher than what we have observed under similar excitation conditions for AlGaN bulk layers and active regions that do not contain NCI regions. It is important to note that these measurement conditions do not generate large densities of photo-generated carriers that might lead to anomalously high IQE due to the saturation of defects at room temperature and the reduction of absorption associated with bleaching at low temperature.

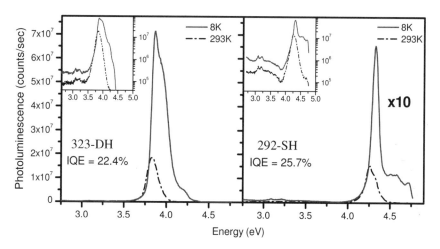

Fig. 1. 8K and 293K cw-PL for 323 nm-DH (left) and 292 nm-SH (right).

The room temperature PL spectra for the SH (right) is also dominated by an intense single peak emission, but at 292 nm, consistent with the larger AlN mole fraction in the active region. The PL spectra measured at 8K again shows about a factor of 4 increase in the intensity of the peak emission and a considerably wider high energy shoulder. The high energy shoulder is attributed to emission from the thick MOCVD grown AlGaN template that was used a substrate for this structure, as well as emission from the hetero-barrier and band edge of the active region. The IQE for this sample is estimated to be ~ 25.7%, similar to that of the 323 nm-DH despite the significantly higher AlN mole fraction in the structure. However, it is important to note that while both samples have similar IQE, the intensity of the PL emission from the 292 nm-SH sample is approximately 10x weaker than that of the 323 nm-DH at both room temperature and low temperature.

3.2. *Time resolved photoluminescence studies*

To further understand the nature of the reduced PL intensity in the 292 nm-SH despite the high IQE of the NCI regions within the active regions, carrier dynamics in both heterostructures were studied by time-resolved photoluminescence (TRPL). Figure 2 shows the TRPL studies for these structures under similar excitation conditions. The PL lifetime of the 323 nm-DH was determined to be ~ 650 ps, significantly larger than the ~ 500 ps lifetime observed for our high quality PA-MBE GaN deposited on HVPE-grown GaN [7]. In contrast, the PL lifetime of the 292 nm-SH is only ~175 ps, considerably shorter than that observed for the 323 nm-DH but consistent with the 292 nm-SH having reduced cw-PL emission intensity.

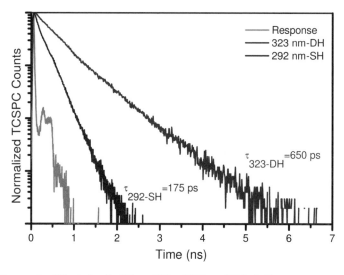

Fig. 2. Time resolved PL studies for 323 nm-DH and 292 nm-SH, including system response limit.

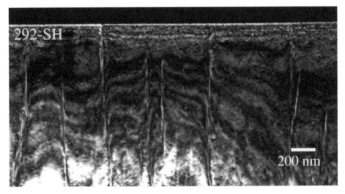

Fig. 3. Cross-sectional TEM images showing PA-MBE/ template interface for 323 nm-DH (top) and 292 nm-SH (bottom).

3.3. *Transmission electron microscopy studies*

To examine the defect structure of the samples, transmission electron microscopy studies were performed on cross-sectional samples of both structures. Figure 3 shows (0002) dark field images of the 323 nm-DH and the 292 nm-SH, examining the heterointerface between the PAMBE growth of the heterostructure on the HVPE or MOCVD template, respectively. It is apparent in both cases that the PA-MBE heteroepitaxy does not generate an appreciable number of new threading dislocations at the heterointerface. However, it is also evident that the 323 nm-DH (top) has over an order of magnitude more threading dislocations than the 292 nm-SH (bottom).

3.4. *290 nm Double Heterostructure Ultraviolet Light Emitting Diodes*

Double heterostructure UVLEDs were realized from the 292 nm-SH by depositing an electron blocking layer and a hole injection layer by MOCVD. These devices were fabricated into flip-chipped LEDs consisting of two interconnected 0.1 mm x 1 mm p-mesas that were hybridized to a double bonded copper (DBC) submount. The commercially available DBC substrate consists of a 25 mil-thick polycrystalline AlN substrate that has a 13 mil-thick copper film bonded to it on each side. Submounts were produced from this substrate by removing the copper film entirely from one side while patterning the film on the opposite side so that it can be employed to inject current into the device. Subsequently, the UVLED chip and the submount were hybridized using standard flip-chip bonding techniques and employing a Au/Sn eutectic as the hybridizing metal. The L-I curves and spectra of these DH-UVLEDs were measured without additional heat management, due to the size of the submount chip, by placing the device chip at the entrance of a calibrated Ocean Optics integrating sphere.

The fabricated devices have a reasonable turn-on voltage of ~6.5V @ 20 mA, but also a high series resistance of ~50 Ω. Figure 4 (right) shows the measured electroluminescence (EL) spectra from a typical device plotted on a logarithmic scale as the cw- drive current is increased from 50 to 260 mA. The EL spectra are dominated by a single peak occurring at 291 nm that agrees well with PL studies of the active region. However the deep level rejection is found to be only ~20x, significantly poorer than what we have previously reported for similar 326 nm devices[5]. There was no observable red-shift in the emission peak with increasing drive current during these measurements, indicating that the DBC submount helps to enhance heat spreading away from the device chip despite the lack of a larger heatsink.

Figure 4 (left) shows the L-I curve for a typical device under cw and pulse testing using a 20% duty cycle and a period of 200 μs. A peak output power of 10 μW was obtained for a cw drive current of 225 mA. A higher output power of 38 μW was obtained under pulse conditions for a drive current of 180 mA. Since the devices were only placed at the entrance, rather than inside, of the integrating sphere due to the size of the submount, the power measurements are somewhat reduced.

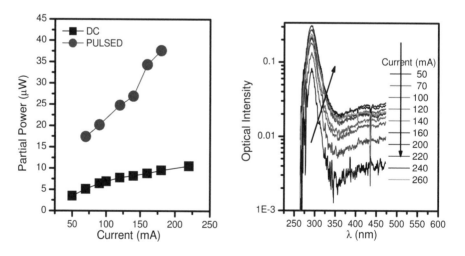

Fig. 4. L-I curve (left) and EL spectra (right) for flip-chipped DH-NCI-UVLED emitting at 292 nm.

4. Discussion

The shorter PL lifetime and reduced room temperature cw-PL intensity observed for the 292 nm-SH are consistent with a larger density of nonradiative centers and concomitant shorter nonradiative lifetime in this material as compared with the 323 nm-DH. However, even allowing for differences in collection efficiency as a function of Al content/wavelength, the much smaller integrated cw-PL intensity for the 292 nm-SH at 8K, where non-radiative processes are negligible, suggests that it possesses a lower density of NCI-regions than the 323 nm-DH. This result implies that there is a greater concentration of carriers in the smaller density of NCI regions in the 292 nm-SH, leading to a shorter room temperature radiative lifetime[4] consistent with the large peak IQE observed, even in the presence of such a high density of nonradiative centers.

Insight into the nature of these nonradiative centers can be found from the TEM data. The presence of threading dislocations or point-defects that decorate such extended defects is widely considered to inhibit radiative recombination, and therefore one might have expected that the 292 nm-SH possessed the greater density of these defects. However, the TEM studies show that the 292 nm-SH has a significantly smaller density of threading dislocations than the 323 nm-DH, suggesting that the nonradiative centers are not associated with threading dislocations. Given the propensity of Al to getter impurities, the shorter nonradiative lifetime at higher Al content implies that the nonradiative centers are point defects. This idea is consistent with previous observations in high quality GaN films grown on 85 μm-thick HVPE GaN templates, where low dislocation density films were found to have diminished PL intensity and carrier lifetime due to the incorporation of a higher density of point defects induced by the growth conditions[7]. Similary, Isono and coworkers attribute the enhancement in the PL intensity they observe in their AlGaN MQWs to the reduction of point defects in their films [8]. Determining the nature of these point defects and how their incorporation can be suppressed requires further study.

The performance of the devices corresponds to a wall plug efficiency, optical power out vs. electrical power in, well below 0.1%. Assuming the calculated extraction efficiency of ~5% and the measured 25.7% IQE for the active region of the device, one would expect and external quantum efficiency closer to ~1% for unity injection efficiency. The much lower wall plug efficiency suggests that carrier injection into the active region is a very serious problem with these devices and is likely related to hole injection, which may suffer due to the formation of defects at the MOCVD-MBE heterointerface.

5. Conclusions

In summary, we observe that NCI-AlGaN active regions emitting at ~290 nm exhibit reduced integrated photoluminescence intensity and PL lifetime relative to 320 nm active regions that have significantly lower AlN mole fraction, despite having more than an order of magnitude fewer threading dislocations. This behavior is attributed to nonradiative recombination associated with the presence of a larger density of point defects in the higher Al content samples. These point defects are ameliorated somewhat by the lower density of NCI AlGaN regions in the higher Al content samples, which leads to a larger concentration of carriers in the NCI and concomitant reduced radiative lifetime that may account for the high observed peak IQE. Double heterostructure, flip-chip mounted, ultraviolet light emitting diodes operating at 292 nm have been demonstrated but show low wall plug efficiencies below 0.1 % despite the high IQE of the active region. This result suggests that the devices suffer from hole injection problems near the heterointerface between the MOCVD-grown hole injection layers and the PA-MBE-grown active region.

6. Acknowledgements

The authors would like to acknowledge Professor A. Khan, University of South Carolina, and Technologies and Devices International, Silver Spring MD, for the MOCVD- and HVPE-grown AlGaN templates employed in this study, respectively.

References

1. Nichia Corporation Specification Sheet for Model NCSU033A (T) UV LED. Catalog No. 061218
2. Sensor Electronic Technology, Inc. UVTOP 280 nm Specification Sheet (www.s-et.com)
3. C. J. Collins, A.V. Sampath, G.A. Garrett, W.L. Sarney, H. Shen, M. Wraback, A.Yu. Nikiforov, G.S. Cargill, III, and V. Dierolf, *Appl. Phys. Lett.* **86**, 31916-31911-3, (2005)
4. M. Wraback, G.A. Garrett, A.V. Sampath, and H. Shen, *Int. J. of High Speed Electron Syst* **17**, 179 (2007)
5. A.V. Sampath, M.L. Reed, G.A. Garrett, E.D. Readinger, H. Shen, M. Wraback, C. Chua, N.M. Johnson, A. Usikov, O. Kovalenkov, L. Shapovalova, and V. Dmitriev", *Phys. Stat. Sol. C,* **5**, 2303 (2008).
6. A.V. Sampath, G. A. Garrett, C. J. Collins, W. L. Sarney, E.D. Readinger, P.G. Newman, H. Shen, and M. Wraback, *J. of Electron. Mat.,* **35**, 641 (2006).
7. A.V. Sampath, G. A. Garrett, C.J Collins, P. Boyd, J.Y. Choe, P.G. Newman, H. Shen, M. Wraback, R.J. Molnar and J. Caissie, *J. of Vac Sci and Tech B* **22**, 1487 (2004).
8. K Isono, E. Niikura, K. Murakawa, F. Hasegawa and H. Kawanishi, *Jpn J. Appl. Phys.,* **46** (9A), 5711 (2007)

International Journal of High Speed Electronics and Systems
Vol. 19, No. 1 (2009) 77–83
© World Scientific Publishing Company

SURFACE ACOUSTIC WAVE PROPAGATION IN GaN-ON-SAPPHIRE UNDER PULSED SUB-BAND ULTRAVIOLET ILLUMINATION

VENKATA S. CHIVUKULA[1], DAUMANTAS CIPLYS[1,2], KAI LIU[1],
MICHAEL S. SHUR[1], REMIS GASKA[3]

[1]*Center for Integrated Electronics, Rensselaer Polytechnic Instutite, 110 8th Street*
Troy, New York 12180, U.S.A
[2]*Department of Radiophysics, Vilnius University, Sauletekio 9*
Vilnius 10222, Lithuania
[3]*Sensor Electronic Technology Inc., 1195 Atlas Road,*
Columbia, SC 29209, U.S.A
chivuv@rpi.edu

We investigated transient amplitude and phase response of GaN-on-sapphire SAW delay-line device subjected to pulsed sub-band UV illumination. We correlated these results with photoluminescence measurements in two GaN samples with the same emission spectra but different carrier lifetime. The SAW response measurements showed that under pulse illumination the sample with shorter lifetime exhibited gradual rise in the phase and amplitude change in millisecond range. Under similar conditions, the sample with a longer lifetime responded faster. We attribute this change to the presence of defect related transitions of the photoexcited carriers and their interaction with surface acoustic waves. Our results demonstrate the possibility of characterization of compensated GaN samples by measuring the phase and amplitude variations in the SAW transmission mode under pulsed UV excitation.

Keywords: Surface Acoustic Wave; Ultraviolet Sensor; Gallium Nitride

1. Introduction

GaN and related III-Nitride semiconductors are very promising materials for applications as surface acoustic wave (SAW) devices, including solar-blind [1] and visible-blind [2,3] ultraviolet sensors. In spite of impressive progress in growing high-quality GaN/sapphire epitaxial substrates, inherent problems such as large leakage currents and lifetime limitations often plague GaN-based optoelectronic devices. Traditionally, optical characterization techniques such as photoluminescence [4], reflectance [5] and Raman scattering [6] are widely used to understand origin and nature of intrinsic defects states in GaN. In this paper we demonstrate that SAW response to UV radiation is correlated to the materials quality characterized by non-equilibrium carrier lifetime.

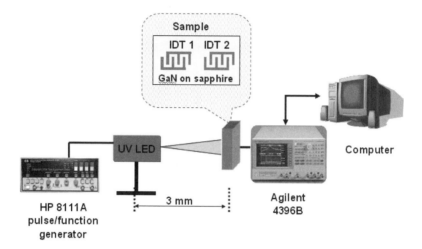

Figure 1: Schematic of the experimental set-up.

We investigated transient amplitude and phase response of SAW delay-line devices in the transmission mode under pulsed UV radiation from LED with a wavelength of 375 nm, which is slightly longer than that of GaN energy gap. Then, we correlated these results with the emission spectra and carrier lifetime obtained using photoluminescence (PL) measurement on both samples.

2. Experiment

Two GaN-on-sapphire samples grown by MOCVD technique were used for SAW device fabrication. Sample A consisted of 2.4 μm GaN layer deposited on c-sapphire substrate. Aluminum-film interdigital transducers (IDTs) with period of 16 μm, aperture 1.3 mm, 90 finger pairs, and 6.5 mm edge-to-edge spacing between the IDTs were fabricated by photolithography technique. The dark value of DC resistance measured between the electrodes was 3.8 GΩ. The SAW of 16 μm wavelength propagated in $\left[1\bar{1}00\right]$ direction of sapphire substrate with the velocity of 4900 m/s. Sample B consisted of GaN layer deposited on c-plane sapphire substrate with aluminum-film IDTs with a period of 24 μm, 100 finger pairs, and 4 mm edge-to-edge spacing fabricated by photolithography. The SAW velocity in this sample was calculated to be 5380 m/s. Such velocities for GaN were earlier reported [7] along the substrate $\left[11\bar{2}0\right]$ direction. The dark value of DC resistance measured between the IDT electrodes in this sample was 109 GΩ. Fig.1 shows the schematic of experimental set-up. The pulse train from HP8111A pulse/function generator switches the 375 nm UV LED manufactured by Nichia Corporation. The external trigger mode of the network analyzer was enabled to synchronize the device sweep with the pulse train and simultaneously record the corresponding changes in phase/amplitude of the SAW delay-line in transmission mode. The data from network analyzer were processed and stored in the computer through GPIB interface.

The time resolved photoluminescence in both GaN samples was excited by the fourth harmonic (266 nm) of the YAG:Nd^{+3} mode-locked laser radiation pulses of energy $h\upsilon =$ 4.66 eV and duration $\tau_L = 30$ ps. The laser spot diameter on the sample was measured to be 90 µm. The detection and data acquisition system in time domain uses the streak camera from Hamamatsu. All the measurements were conducted in dark at room temperature. The details about the PL experimental set-up are described elsewhere [8].

3. Results and Discussion

The phase of the SAW signal at the output transducer is retarded with respect to the input transducer by the amount $\phi = 360^\circ \times fL/V$, where f and V are the SAW frequency and velocity, respectively, and L is the SAW travel distance. It was earlier reported that variation in velocity of the SAW under UV is responsible for the phase and amplitude change [9]. However, the dynamics of phase and amplitude response for pulsed 375 nm UV excitation was not completely investigated. In general, the phase shift $\Delta\phi$ in the output transducer is related to velocity variation ΔV by the following expression:

$$\frac{\Delta\phi}{\phi} = \frac{\Delta V}{V} \tag{2}$$

Here, it is assumed that the total SAW path is homogenously illuminated by UV. The change under UV illumination of the signal amplitude at the output IDT can be attributed to the dependence of SAW attenuation on the sheet resistivity of GaN layer, which can be expressed as [10]

$$A = 8.68\pi K^2 \frac{f}{V} \frac{\varepsilon_0 \varepsilon V R_S}{1 + \left(\varepsilon\varepsilon_0 V R_S\right)^2} \tag{3}$$

where K^2 is the electromechanical coupling coefficient for the SAW in a given structure, ε_o is the dielectric permittivity of free space, and ε is the permittivity of the material under consideration. In our experiments, the SAW phase response at the synchronism frequency f_o was measured as a function of time. The phase change of 7.5° and the amplitude change of 0.6 dB were observed in sample A for pulse generator voltage of 3.5 V and 4 ms pulse duration as shown in Fig. 2 (a and b). The phase and amplitude shifts were observed to be negative with reference to the baseline value. This happens because carriers generated in the sample under UV excitation leads to the screening of the SAW piezoelectric fields and therefore to the SAW velocity decrease, which manifests as the negative shift in phase responses. The photoconductivity is a cause for the SAW attenuation increase, which manifests as the decrease in SAW amplitude. For our experimental parameters, V=4900 m/s, $f = 307$ MHz and L= 6.5 mm, one obtains from Eq. (2) the relative change in velocity $|\Delta V/V| = 5.1\times10^{-5}$ at a UV power density of 1.46 µW/mm^2 (assuming the area of 4×6.5 mm^2 between the IDT region is illuminated) or defining the device responsivity S as a relative velocity change per unit power density, $S = 3.5\times10^{-5}$ (µW/mm^2)$^{-1}$.

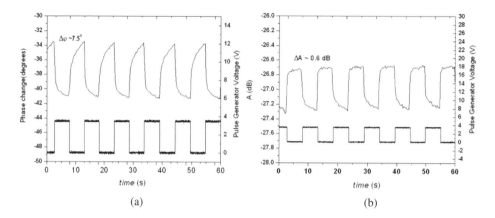

Figure 2: (a) Time dependent pulsed UV induced change in phase of output SAW transducer signal in sample A (b) corresponding amplitude change.

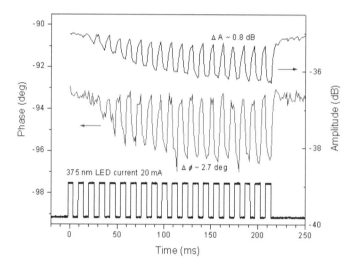

Figure 3: Time dependent pulsed UV induced change in phase and amplitude of output SAW transducer signal in sample B.

Using the same set-up shown in Fig. 1, we performed the similar investigations of the SAW phase and amplitude variations under pulsed 375 nm UV LED in sample B. The essential difference we observed for this sample was that neither phase nor amplitude of the SAW exhibited any changes during a few initial pulses of the LED-driving voltage pulse train. The SAW phase and amplitude changes under pulse UV illumination gradually increased with pulse number of the train, in contrast to sample A, where these changes were practically the same for all the pulses of a train. As one can see from Fig. 3, the phase and amplitude changes attains maximum values of $2.7°$ and 0.8 dB, respectively, after about 100 ms at voltage 8 V and duration 4 ms of the LED driving pulses. The relative change in velocity $|\Delta V/V| = 4.5 \times 10^{-5}$ corresponding to maximum phase change was calculated using Eq. (2) for UV power density of 106 μW/mm^2 (assuming the area of 4×2 mm^2 is illuminated). Evaluation of the responsivity for this sample yields the value $S = 4.2 \times 10^{-7}$ (μW/mm^2)$^{-1}$, which is considerably lower than that for sample A.

We also observed that repeated measurements of amplitude/phase on the same sample by LED switching led to the decrease in magnitude of phase and amplitude variations in both the samples, which can be attributed to persistent photoconductivity earlier reported in GaN [11].

The processes of photoconductivity causing the changes in SAW propagation parameters are strongly dependent on the wavelength of the exciting light and the GaN semiconductor properties. Since the GaN absorption band edge is around 362 nm, one can expect that band-to-band excitations are not important and mainly transitions involving band-tails and localized states are responsible for the photoconductivity at 375 nm. Photoconductive spectroscopy [12] and photothermal deflection spectroscopy [13] studies on epitaxial GaN grown by MOCVD technique indicate broad distribution of defect related states within the forbidden gap of GaN, which act as traps for recombination of generated photocarriers. However these studies do not reveal information about carrier lifetime and emission spectra, which is important in understanding the interaction of SAW with photoexcited carriers. For this purpose, photoluminescence measurements were performed using the fourth harmonic (266 nm) of YAG:Nd^{+3} laser. They revealed that the two samples have similar emission spectra with the peak PL intensity around 3.4 eV corresponding to GaN, but different non-equilibrium carrier lifetimes (as shown in Figs. 4 and 5).

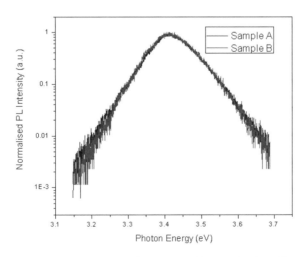

Figure 4: Normalized PL emission spectra in GaN-on-sapphire samples A and B at excitation power density of 110 MW/cm^2.

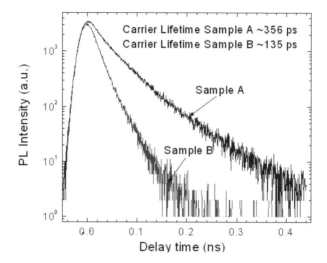

Figure 5: PL intensity dependence on time delay for GaN-on-sapphire samples A and B at excitation power density of 110 MW/cm².

The carrier lifetimes of 356 ps and 135 ps for sample A and B, respectively, were extracted from the slopes of exponential decay in PL intensity with time shown in Fig. 4 and analytically expressed by the relation $I(t) = I_0 \exp(-t/\tau)$, where τ is the carrier lifetime. It has been found from literature [14] that the lifetime of neutral donor bound exciton in n-GaN usually lies in the range from 35-530 ps. The shorter carrier lifetime of 135 ps found in sample B can be attributed to its higher dislocation density. These results can be correlated to our observation of the SAW phase and amplitude behavior in sample B under pulsed UV illumination. We speculate that the higher concentration of photocarriers with longer carrier lifetime under UV illumination in sample A is responsible for enhanced response in SAW phase and amplitude. Meanwhile, for the sample B having lower carrier lifetime, the extent of increase in carrier lifetime by propagating surface acoustic wave is smaller, which leads to more gradual and smaller changes in phase and amplitude responses.

   In conclusion, our results demonstrate the possibility of characterization of compensated GaN samples by measuring the phase and amplitude variations in the SAW transmission mode under pulsed UV excitation.

Acknowledgement

The work at RPI has been supported by the National Science Foundation under the auspices of I/UCRC "Connection One."

References

1. D. Ciplys, R. Rimeika, M.S. Shur, A. Sereika, R. Gaska, J. W. Yang, and M.A. Khan, Radio frequency response of GaN-based SAW oscillator to UV illumination by the Sun and man-made source, *Electronic Letters*, **38**(3), 134-135 (2002).

2. D. Ciplys, R. Rimeika, M.S. Shur, S. Rumyantsev, Visible-blind photoresponse of GaN-based surface acoustic wave oscillator, *Applied Physics Letters*, **80**(11), 2020-2022 (2002).

3. D. Ciplys, R. Rimeika, M.S. Shur, N. Pala, A. Sereika, R. Gaska, Q. Fareed, Ultraviolet-sensitive AlGaN-based Surface Acoustic Wave Devices, IEEE SENSORS 2004, Vienna, Austria, Oct. 24-27 (2004).

4. M. Smith, G.D. Chen, J. Z. Li, J. Y. Lin, H. X. Jiang, M. Asif Khan, C. J. Sun, Q. Chen, J. W. Yang, Free excitonic transitions in GaN, grown by metal-organic chemical-vapor deposition, J. Appl. Phys. **79**(9), 7001 (1996).

5. R. Dingle, D. D. Sell, S. E. Stokowski, and M. llegems, Absorption, Reflectance, and Luminescence of GaN Epitaxial Layers, *Phys. Rev. B*, **4**, 1211-1218 (1971).

6. V. Yu. Davydov et al., Phonon dispersion and Raman scattering in hexagonal GaN and AIN, *Phys. Rev. B*, **58**, 12899-12907 (1998).

7. R. Rimeika, D. Ciplys, M. S. Shur, R. Gaska, M. A. Khan, J. Yang, Electromechanical Coupling Coefficient for Surface Acoustic Waves in GaN-on-Sapphire, *phys. stat. sol. (b)* **234**(3), 897-900 (2002).

8. E. Kuokstis, G. Tamulaitis, K. Lui, M. S. Shur, J. W. Li, J. W. Yang, and M. Asif Khan, Photoluminescence dynamics in highly nonhomogeneously excited GaN, *Appl. Phys. Lett.*, **90**(16), 161920 (2007).

9. D. Ciplys, R. Rimeika, M.S. Shur, J. Sinius, R. Gaska, Yu. Bilenko, Q. Fareed, UV-LED controlled GaN-based SAW phase shifter, *Electronic Letters,* **42** (21), 1254-1255 (2006).

10. R. Adler, Simple Theory of Acoustic Amplification, IEEE Trans. Son. Ultrason. **SU-18**(3), 115-118 (1971).

11. C. H. Qiu, and J. I. Pankove, Deep levels and persistent photoconductivity in GaN thin films, Appl. Phys. Lett. **70**(15), 1983-1985 (1997).

12. C. H. Qiu, C. Hoggatt, W. Melton, M. W. Leksono, and J. I. Pankove, Study of defect states in GaN films by photoconductivity measurement, Appl. Phys. Lett. **66** (20), 2712-2714 (1995).

13. O. Ambacher, W. Rieger, P. Ansmann, H. Angerer, T. D. Moustakas, and M. Stutzmann, Sub-bandgap absorption of gallium nitride determined by Photothermal Deflection Spectroscopy, Solid State Communications, **97**(5), 365-370 (1996).

14. M. O. Manasreh and H. X. Jiang (editors), III-Nitride Semiconductor Optical Properties I, Vol. 13, (Taylor & Francis, New York, 2002).

International Journal of High Speed Electronics and Systems
Vol. 19, No. 1 (2009) 85–92
© World Scientific Publishing Company

SOLAR-BLIND SINGLE-PHOTON 4H-SiC AVALANCHE PHOTODIODES

ALEXEY VERT

General Electric Global Research Center,
Niskayuna, NY 12309, USA
vertiatc@research.ge.com

STANSILAV SOLOVIEV, JODY FRONHEISER, PETER SANDVIK

General Electric Global Research Center,
Niskayuna, NY 12309, USA

A solar blind 4H-SiC single photon avalanche diode (SPAD) is reported. The SPAD with separate absorption and multiplication layers was designed for operation with low dark counts. A thin film optical filter deposited on a sapphire window of the device package provided sensitivity in the wavelength range between 240 and 280 nm with a very high solar photon rejection ratio. An estimated dark current of 0.4 pA (0.75 nA/cm2) at a gain of 1000 was measured on a device with an effective mesa diameter of 260 μm. A single photon detection efficiency of 9% (linear mode) and 9.5% (gated Geiger mode) were achieved at a wavelength of 266 nm for the same device. Corresponding dark count rate and dark count probability were 600 Hz and 4×10^{-4}.

Keywords: Avalanche photodiodes; Single photon detector; Solar-blind detector

1. Introduction

Silicon carbide (SiC) offers many attractive material properties for fabricating compact and reliable single photon ultraviolet avalanche photodiodes (APDs). The detection of very faint signals with high signal to noise ratio in the solar blind wavelength range is desired for various applications, including the corona discharge and flame detection, UV astronomy, biological and chemical detection, detection of jet engines and missile plumes [1,2]. SiC APDs are a promising alternative to replace photomultiplier tubes (PMTs) when ruggedness, compactness and low cost are needed.

Commercially available Si avalanche photodiodes demonstrate moderate quantum efficiencies in the solar-blind range, but require expensive optical filters to achieve a high solar photon rejection ratio as their response extends through the visible wavelength range. Si-based APDs with very low dark counts were demonstrated when they were cooled to liquid nitrogen temperatures [3]. However, typical dark current densities of these devices at a gain of more than 10 are on the order of ~100 nA/cm^2 at room temperature, which makes them hard to operate in a single-photon detection mode without cooling.

GaN-based APDs were demonstrated with a high sensitivity in the solar-blind region and gain of more than 1000 [4,5,6]. Reported dark currents at a gain of 1000 were typically more than 100 nA/cm² for these devices likely due to high defect densities present in the GaN material.

SiC-based APDs have been demonstrated with high quantum efficiencies in the solar-blind wavelength range, and have been shown to have high gain and low dark currents [7,8]. Recently reported devices with dark current densities of 63 nA/cm², demonstrated a gain of more than 10,000 and also showed a very low noise equivalent power of 20 fW [9].

Typical SiC-based photo detectors show response in the range from 200 to 400 nm with a peak value between 270 and 300 nm wavelengths. In order to limit the SiC APD response to the solar-blind wavelengths, an optical filter may be used to reject the light above approximately 280 nm, where some light absorption occurs due to indirect nature of SiC band structure.

2. Device Fabrication and Packaging

The devices were fabricated using epitaxial layers on 3-inch diameter n-doped 4H-SiC substrate purchased from CREE, Inc. (Durham, NC). The photodiode separate absorption and multiplication (SAM) structure had effective p-n junction areas of approximately 260 μm. The light sensitive area was at least 200 μm in diameter. The structure consisted of a 0.2 μm n⁺ cap layer (N_d=2x10^{18} cm^{-3}), a 2.0 μm n⁻-layer (N_d=1.0x10^{16} cm^{-3}), a 0.45 μm n-layer (N_d=6x10^{17} cm^{-3}), a 2.7 μm n⁻-layer (N_d=1.0x10^{16} cm^{-3}), and a 2 μm p⁺-layer (N_A=10^{18} cm^{-3}) as shown in Fig. 1. The devices were designed as a non-reach-through structure, i.e. an electric field was completely terminated in the n-layer and did not extend in the n⁻ absorption layer. Since there was no electric field in the absorption region, the photo-generated carriers traveled via diffusion to the multiplication region.

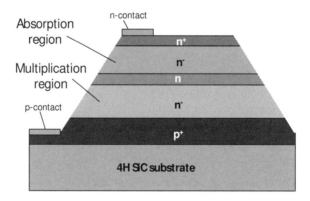

Fig. 1. Device cross-section showing the separate absorption and multiplication regions.

The processing steps used to fabricate the devices included mesa etching, passivation oxide deposition, cathode and anode contact formation, and passivation oxide removal in the active area. Inductively coupled plasma reactive ion etching was used to define the mesas with the beveled geometry in order to eliminate edge breakdown effects. A thermally grown SiO_2 film was deposited to passivate the surface. Ni was sputtered to form the n-type contacts and a multiple-layer stack of Ti/Al/Ti/Ni was deposited to form the p-type contacts. Both contacts were annealed simultaneously at 1050^0C in a N_2 ambient. Finally, Ti/Au metal pads were deposited on top of the ohmic contacts and the passivation layer was removed in the active region by wet etching.

Prior to packaging the devices, an optical filter comprised of thin dielectric layers was deposited on a sapphire window by Barr Associates Inc. [10]. The devices were attached to the header using non-conductive epoxy and connected to the package leads using wire-bonds. The cap with the filter window was attached to protect the package. The inset of Fig. 2 shows the photograph of the TO8 header package of the device.

3. Device Characterization

Fig. 2 shows the responsivity of the fabricated APD devices with and without the filter. Quantum efficiency was measured in the wavelengths range from 200 to 1000 nm at a reverse bias of 50 V in the linear regime at unity gain.

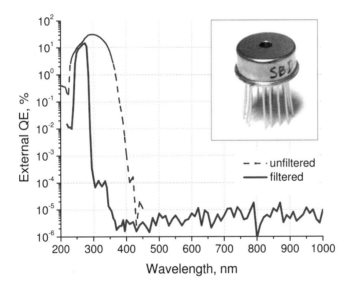

Fig. 2. Measured external quantum efficiency of 4H-SiC SAM APDs with and without the filter. Inset: photograph of the device package.

The measured gain, dark current and photocurrent characteristics are shown in Fig. 3. The filtered APD device demonstrated an estimated dark current density of 0.75 nA/cm^2 at a gain of 1000 with an external quantum efficiency peak of 15% near 280 nm. The low dark current allowed operating fabricated SAM APDs in the linear "quasi-Geiger" and Geiger modes without any damage or degradation of the devices.

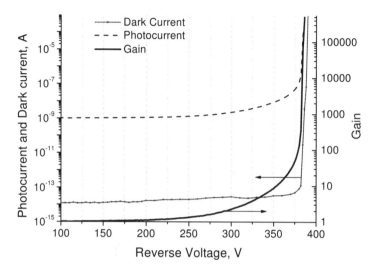

Fig. 3. Reverse-bias I–V characteristics of SiC SAM APD with the effective mesa diameter of 260 μm.

The breakdown voltage was measured to be around 385 V and the bias voltage for the linear "quasi-Geiger" mode measurements was swept between 385 V and 388 V. The response of the APD device in this regime to a laser pulse was a sharp avalanche current spike, which was passively quenched by a series resistor without any after-pulsing effects observed. An experimental setup utilized for the passive quenching linear mode measurements is shown in Fig. 4. A 266nm laser beam with an average of 0.5 photons during a 500-ps pulse and an 8,300-Hz repetition rate was focused within the device. The bias voltage was scanned and the single photon detection efficiency (SPDE) and dark count rate (DCR) were measured for the same bias to obtain the curve shown in Fig. 5.

The detection probability was estimated by normalizing the number of avalanche pulses to the total number of laser pulses and the dark count rate was measured by counting the total number of avalanche pulses in the dark during one-second. The single photon detection efficiency was nearly 9% and a dark count rate of 600 Hz was measured at a 388 V bias. There was a noticeable occurrence of after-pulsing at biases above 388 V, which suggested that passive quenching may not be adequate for higher gains.

For gated passive quenching Geiger mode measurements, the dc bias voltage was set fixed at 375 V, or 10 V below the breakdown voltage. A schematic of the measurement setup is shown in Fig. 6. The excess ac-coupled voltage above the dc voltage was applied during a 9-ns gate pulse. A 266-nm laser beam similar to one used in the linear mode measurement was synchronized with the gate pulse. The avalanche pulses were discriminated and counted within a 2-ns window. Fig. 7 shows the single photon detection efficiency (SPDE) and the dark count probability (DCP). The excess voltage was scanned in the range from 10 to 50 V and the SPDE and DCP were measured for the same over-bias to obtain the curve. A single photon detection efficiency of 9.5% and a dark count probability of 4×10^{-4} were measured at a maximum over-bias.

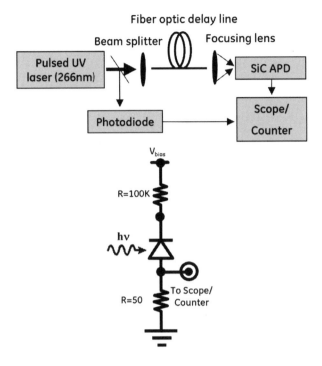

Fig. 4. Block diagram and schematics of passive quenching linear mode setup.

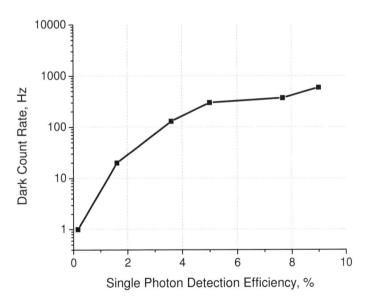

Fig. 5. Linear mode dark count rate versus single photon detection efficiency at 266nm.

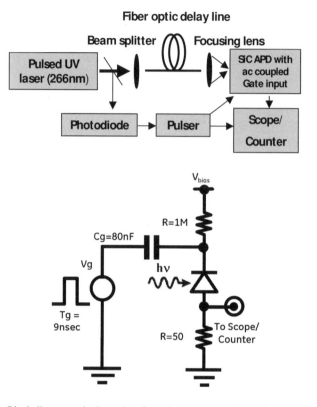

Fig. 6. Block diagram and schematics of gated passive quenching Geiger-mode setup.

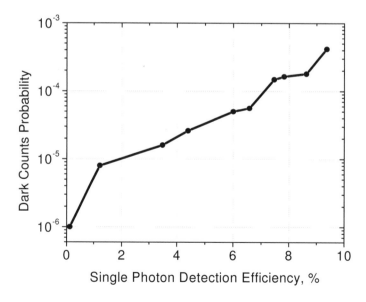

Fig. 7. Geiger mode dark counts probability versus single photon detection efficiency at 266nm.

Gated Geiger-mode regime operation has achieved higher gains and better single photon detection efficiency, but un-gated linear mode measurements have also demonstrated promising results. The simplicity to implement passive quenching is one of the attractive advantages of the linear mode measurements.

4. Summary

We report solar-blind low dark current SAM 4H-SiC APDs capable operating in both linear "quasi-Geiger" and Geiger modes. The peak 15% quantum efficiency was demonstrated at 280 nm wavelength with a very high solar photon rejection ratios. Single photon detection efficiencies of 9% and 9.5% were achieved at 266nm wavelength for un-gated linear mode and gated Geiger-mode respectively. Low dark count rates were achieved due to estimated low dark current of 0.4 pA at a gain of 1000 with a 260 μm diameter device. Absence of defects in the device structure such as dislocations and micro-pipes and uniform spatial distribution of the multiplication gain are believed to be key factors for efficient single photon detection and low dark count rate observed in studied large-size APDs.

5. Acknowledgments

The authors are grateful to Professor J. Campbell for fruitful discussions. The authors also acknowledge Dr. H. Temkin for program management of this effort under the DARPA Deep Ultraviolet Avalanche Photodiode Program. This publication was prepared with support of U.S. Army Award Number W911NF-06-C-0160. However, any opinions, findings, conclusions or other recommendations expressed herein are those of the authors and do not necessarily reflect the views of U.S. Army or DARPA.

References

1. P. Schreiber, T. Dang, T. Pickenpaugh, G. Smith, P. Gehred, C. Litton, Solar blind UV region and UV detector development objectives, *Proc. of SPIE - The International Society for Optical Engineering*, vol. 3629, 230-248 (1999).
2. D. Engelhaupt, P. Reardon, L. Blackwell, L. Warden, B. Ramsey, Autonomous long-range open area fire detection and reporting, *Proc. of SPIE - The International Society for Optical Engineering*, vol. 5782, *Thermosense XXVII*, 164-175 (2005).
3. T. Isoshima, et al., Ultrahigh sensitivity single-photon detector using a Si avalanche photodiode for the measurement of ultraweak biochemiluminescence, *Rev. Sci. Instrum.*, vol.66, no. 4, 2922–2926 (1995).
4. R.D. Dupuis, J.B. Limb, D. Yoo, Y. Zhang, J.H. Ryou, S.C. Schen, GaN ultraviolet avalanche photodiodes grown on 6H-SiC substrates with SiN passivation, *Electron. Lett.*, vol. 44, no. 4 (2008).
5. S. Verghese, K.A. McIntosh, R.J. Molnar, L.J. Mahoney, R.L. Agrawal, M.W.Geis, K.M. Molvar, E.K. Duerr, and I. Mengalis, GaN avalanche photodiodes operating in linear-gain mode and Geiger mode, *IEEE Trans. Electron Devices*, vol. 48, 502–511 (2001).

6. J.L. Pau, R. Mcclintock, K. Minder, C. Bayram, P. Kung, M. Razeghi, Geiger-mode operation of back-illuminated GaN avalanche photodiodes, *Appl. Phys. Lett.*, vol. 91, no. 4, 041104 (2007).

7. X. Xin, F. Yan, X. Sun, P. Alexandrove, C.M. Stahle, J. Hu, M. Matsumara, X. Li, M. Weiner, and H.J. Zhao, Demonstration of 4H-SiC UV single photon counting avalanche photodiodes, *Electron. Lett.*, vol. 41, no. 4, pp. 212-214 (2005).

8. X. Guo, A.L. Beck, Z. Huang, N. Duan, J.C. Campbell, D. Emerson, J.J. Sumarekis, Performance of Low-Dark-Current 4H-SiC Avalanche Photodiodes With Thin Multiplication Layer, *IEEE Trans. Electron Devices*, Sept. 2006, vol. 53, no. 9, 2259-2265 (2006).

9. X. Bai, X. Gou, D.C.McIntosh, H.Liu, J.C. Campbell, High Detection Sensitivity of Ultraviolet 4H-SiC Avalanche Photodiodes, *IEEE J. Quantum El.*, vol. 43, no. 12, 1159-1162 (2007).

10. Barr Associates, Inc. Westford, MA 01886 USA.

International Journal of High Speed Electronics and Systems
Vol. 19, No. 1 (2009) 93–100
© World Scientific Publishing Company

MONTE CARLO SIMULATIONS OF In$_{0.75}$Ga$_{0.25}$As MOSFETs AT 0.5 V SUPPLY VOLTAGE FOR HIGH-PERFORMANCE CMOS

J. S. AYUBI-MOAK[†], K. KALNA, and A. ASENOV

Device Modelling Group, Department of Electronics & Electrical Engineering, University of Glasgow,
Glasgow G12 8LT, Scotland, United Kingdom
[†]*jasons@elec.gla.ac.uk*

The performance of implant-free (IF), *n*-type III-V MOSFETs with an In$_{0.75}$Ga$_{0.25}$As channel have been evaluated using a 2D finite-element Monte Carlo device simulator. We investigate the device performance of a set of scaled transistors with gate lengths of 30, 20 and 15 nm at a drain bias of 0.5 V to determine whether this novel architecture can deliver the high drain current at low bias conditions required for high-performance CMOS applications.

Keywords: III-V MOSFETs; implant-free; enhancement-mode MOSFET; Monte Carlo device simulations

1. Introduction

The use of higher mobility compound semiconductors such as InGaAs for *n*-type MOSFETs is now an accepted option for overcoming the performance bottleneck of future silicon CMOS scaling. A very low electron effective mass in the lowest energy valley of III-V materials assures a very high injection velocity, which, in combination with high mobility and low backscattering, promises very high device performance[1]. When combined with the recent development of a high-κ dielectric for GaAs with a low interface state density[2,3], the development of a suitable and manufacturable III-V MOSFET architecture able to sustain continued downscaling beyond the 22 nm node is becoming more realistic[4,5].

Introducing such novel materials into a MOSFET structure, however, requires new design concepts that can take full advantage of higher mobility channel while at the same time maintaining a significant performance advantage over their Si-based counterparts through continued downscaling. One such recently developed architecture that appears to satisfy this criteria is the enhancement-mode, implant-free (IF) MOSFET[6,7]. This design, which does not require implanted source/drain regions or extensions, can take full advantage of the high injection velocity inherent to III-V materials since it does not rely upon surface inversion to achieve enhancement-mode operation. Rather, it utilizes an undoped channel and a single type of doping, e.g., *n*-type doping for NMOSFETs. Furthermore, band-to-band tunneling, which can significantly degrade the performance of devices with implanted source/drain regions, can be greatly reduced as a result of smaller channel thickness and the use of wider bandgap materials surrounding the III-V channel in the IF design architecture.

In this work, using state-of-the-art ensemble Monte Carlo (MC) device simulations, we have investigated the potential device performance of *n*-type IF $In_{0.75}Ga_{0.25}As$ channel MOSFETs with high-κ dielectric scaled to gate lengths of 30, 20 and 15 nm. In the next section, we provide a brief overview of the Monte Carlo simulation approach used in this study and some details of the transport physics included within the modeling tool. In section three, the simulated device structure is described in more detail with device and transfer characteristics presented and discussed in section four. Finally, concluding remarks are offered in section five.

2. Monte Carlo Simulation Approach

The two-dimensional (2D) simulations of scaled implant-free $In_{0.75}Ga_{0.25}As$ channel MOSFETs discussed in the next few sections have been carried out using a heterostructure finite-element MC device simulator MC/MOS[8,9]. This particle-based simulator includes electron scattering with polar optical phonons, inter- and intravalley optical phonons, nonpolar optical phonons, acoustic phonons, and ionized impurities. In addition, the effect of alloy scattering, the impact of strain on the bandgap, electron effective mass, optical phonon deformation potential, and optical phonon energy in the InGaAs channel are also included in the simulation model. This simulation tool has been previously used in the study and design of high-electron mobility transistors (HEMTs) and has been carefully calibrated against experimental *I-V* characteristics of fabricated 120 nm gate length pseudomorphic HEMTs[9]. It has also been used to investigate electron transport in a 120 nm gate length, double δ-doped InGaAs/InAlAs lattice-matched HEMT and demonstrated excellent agreement with measured *I-V* characteristics of both conventional and self-aligned devices[10]. Most recently, this simulator has been calibrated against measured characteristics of a 50 nm gate length InP HEMT with a high indium content ($In_{0.7}Ga_{0.3}As$) channel[11]. These previous studies have validated the transport model currently used within our simulator and its ability to describe highly nonequilibrium electron transport in nano-scale ultrafast devices with high-mobility channels.

Our simulator has also been modified to include the effects of quantum confinement. Quantum corrections are incorporated into the simulator by adopting a potential-smoothing technique called the effective quantum potential (EQP) approach[12]. Using this technique, the classical potential P at each position in real space $\mathbf{r}$ undergoes a convolution with a Gaussian distribution G, to obtain an EQP, P_{eff}, using the following relationships:

$$P_{eff}(\mathbf{r}) = \int d\mathbf{r}\, P(\mathbf{r}+\mathbf{r}')G(\mathbf{r}'),$$

$$G(x) = \frac{1}{a\sqrt{2\pi}}\exp\left(-\frac{x^2}{2a^2}\right) \tag{1}$$

where a is a smoothing parameter. The resulting EQP is then used in the transport engine to propagate the MC particles throughout the simulation space. The corresponding values of a were obtained by matching results of a simple, self-consistent one-dimensional (1D) Poisson-Schrodinger solution of the relevant layer structures in the subthreshold region leading to a = 2.0, 1.85 and 1.85 nm for the 30, 20 and 15 nm gate length devices, respectively.

The EQP approach approximates a shift in the lowest subband in the 2D quantum well (QW) relevant to carrier transport and the shape of the electron distribution[11]. In other words, it moves the electron density centroid away from the semiconductor and dielectric interfaces thus correctly shifting the threshold voltage and decreasing the gate-to-channel capacitances. These effects combined also reduce the saturation current[13,14]. It is important to point out here that this quantum correction approach is simply a numerical smoothing technique applied to the electrostatic potential. We use this approach within the framework of our semi-classical, particle-based model because it correctly approximates the shift in carrier density away from the semiconductor and dielectric interfaces that occur physically within the QW of the actual device in practice. We do not specifically model the quantum confinement within the high-mobility channel, but rather approximate its effect on the resulting carrier transport through a smoothed and shifted electrostatic potential in the high-mobility channel of the device.

3. Implant-Free In₀.₇₅Ga₀.₂₅As MOSFET

A layout of the simulated 2D implant-free In₀.₇₅Ga₀.₂₅As MOSFET is illustrated in Fig. 1. The high-κ dielectric is assumed to be $Ga_2O_3/(Gd_xGa_{1-x})_2O_3$ (GGO) with a static dielectric constant of 20[15]. In order to maintain electrostatic integrity in the 30, 20 and 15 nm gate length structures, the oxide thickness is scaled to 3.0, 2.0 and 1.5 nm, respectively. In

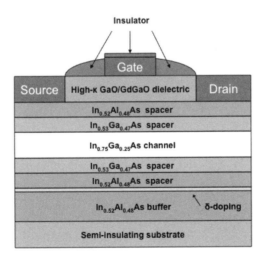

Fig. 1 Layout of the implant-free In₀.₇₅Ga₀.₂₅As MOSFET structure used in the 2D Monte Carlo simulations.

each case, the $In_{0.75}Ga_{0.25}As$ channel thickness is fixed at 5 nm. Total layer thicknesses for each case are summarized in Table 1. An active δ-doping layer concentration of 3×10^{12} cm^{-2} and thickness of 0.5 nm are also held constant in each of the scaled transistors. The workfunction of the metal gate was fixed at 5.3 eV for each device. This parameter is, however, adjusted later to fit the obtained simulation results for an enhancement mode threshold voltage of 0.2 V. It is also important to point out here that the predicted performance characteristics presented and discussed in the next section only reflect simulations of the intrinsic device structures. Therefore, the impact of external contact resistances and other parasitic parameters have been neglected in this study.

Table 1. Layout dimensions for the scaled, implant-free, $In_{0.75}Ga_{0.25}As$ MOSFETs

Thickness of [nm]	Gate Length [nm]		
	30	20	15
High-κ dielectric	3	2	1.5
$In_{0.52}Al_{0.48}As$ spacer	2	1	1
$In_{0.53}Ga_{0.47}As$ spacer	1	0.5	0.5
$In_{0.75}Ga_{0.25}As$ channel	5	5	5
$In_{0.53}Ga_{0.47}As$ spacer	1	1	1
$In_{0.52}Al_{0.48}As$ spacer	2	2	2
δ-doping	0.5	0.5	0.5
$In_{0.52}Al_{0.48}As$ buffer	10	10	10
Semi-insulating substrate	500	500	500

4. Simulation Results

The I_D-V_G transfer characteristics for the 30, 20 and 15 nm gate length device at V_D = 0.05 V and V_D = 0.5 V are shown in Fig. 2 as a function of the difference between gate voltage and threshold voltage and reveal corresponding drive currents of 1360 µA/µm, 1600 µA/µm and 1760 µA/µm, respectively, at V_D = 0.5 V and V_G = 0.5 V. For each gate length, the corresponding threshold voltage of V_T = 0.2 V has been determined via linear extrapolation by adjusting the metal gate workfunction to ϕ_m.= 5.0 eV The simulated devices show continuously improving drive current with scaling which is more than two times higher compared to that expected of equivalent channel length Si MOSFETs at similar bias conditions. The transconductance for the V_D = 0.5 V case is shown in Fig. 3 and reveals extremely high corresponding peak values of 3840 µS/µm, 5100 µS/µm and 5690 µS/µm, respectively. The full set of I_D-V_D characteristic curves are plotted in Fig. 4 for V_G = 0.1, 0.3, 0.5, and 0.7 V. Source and drain contacts for each gate length device are considered to be self-aligned with symmetric source-gate and gate-drain spacings of 30, 20 and 15 nm respectively.

Average electron velocity and energy along the $In_{0.75}Ga_{0.25}As$ channel are plotted in Fig. 5 at V_G-V_T = 0.3 V and V_D = 0.5 V for each of the scaled devices. The position of the arrows indicates the edge of the gate contact at the source end of the device with the drain end fixed at zero for each case. The behavior of the average velocity demonstrates that electrons are rapidly accelerated almost immediately after injection into the channel from the access regions. Upon entering the region beneath the gate, electrons are subjected to the high fringing electric fields that exist there. Subsequently, their transport behavior quickly becomes highly non-equilibrium. Peak velocities of 3.9×10^7 cm/s, 4.9×10^7 cm/s and 5.1×10^7 cm/s near the drain edge of the gate are extracted for each of the

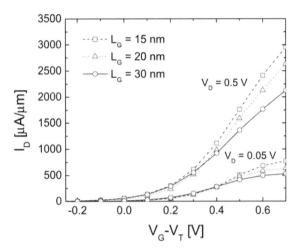

Fig. 2 Transfer characteristics of the 15, 20 and 30 nm gate devices at $V_D = 0.05$ V and $V_D = 0.5$ V. The threshold voltage of 0.2 V has been determined from linear extrapolation by adjusting the metal gate workfunction to 5.0 eV.

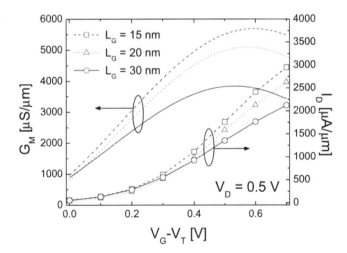

Fig. 3 Transfer characteristics for the 15, 20 and 30 nm gate length devices at $V_D = 0.5$ V. Peak transconductance of 3840 µS/µm, 5100 µS/µm and 5690 µS/µm occur at 0.53 V, 0.58 V and 0.6 V respectively.

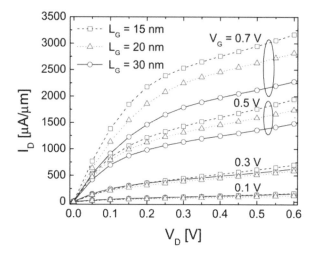

Fig. 4 I_D-V_D characteristics for the 15, 20 and 30 nm gate length devices at V_G = 0.1, 0.3, 0.5 and 0.7 V. The threshold voltage of 0.2 V has been determined from linear extrapolation by adjusting the metal gate workfunction to 5.0 eV.

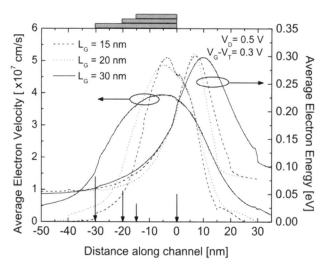

Fig. 5 Average electron velocity and energy along the $In_{0.75}Ga_{0.25}As$ channel of the scaled, implant-free MOSFETs at V_G-V_T = 0.3 V and V_D = 0.5 V. The beginning of the gate is indicated by the arrows with the end of the gate always fixed at zero. The gate length is also illustrated at the top of the plot.

corresponding simulations. After traversing the high field regions underneath the gate, electrons quickly lose energy due to enhanced phonon scattering and frequent transitions to the upper L and X valleys which have larger electron effective masses. Average velocity begins to drop off quickly in each case, but does remain *relatively* high upon entering the drain contact during the scaling process. It is interesting to point out the relative difference in the peak velocity of the 20 nm and 15 nm gate devices compared with that observed when scaling from 30 nm down to 20 nm. This difference in the peak velocities is attributed to extremely high fringing fields at the source end of the 15 nm gate, which facilitates electron transfer to higher effective mass energy valleys in In$_{0.75}$Ga$_{0.25}$As.

5. Conclusions

Using state-of-the-art MC device simulations, we have conducted a preliminary study of the behavior of enhancement-mode, implant-free, In$_{0.75}$Ga$_{0.25}$As MOSFETs assuming a low supply voltage of 0.5 V. We have demonstrated that this recently proposed device technology can in fact deliver very high drive currents of 1360 µA/µm, 1600 µA/µm and 1760 µA/µm, respectively, at this low, 0.5 V supply voltage when fully scaled from a gate length of 30 nm to 20 nm and 15 nm. In addition, we have shown that correspondingly high peak transconductance of 3840 µS/µm, 5100 µS/µm and 5690 µS/µm, respectively, can be achieved for the same device architecture in the scaling process. These simulated devices show continuously improving drive current with scaling which is more than two times higher compared to that expected of equivalent channel length Si MOSFETs at similar bias conditions. The remarkably high drive current at relatively low drain bias makes these devices excellent candidates for future low-power, high-performance CMOS circuits and applications because power dissipation will be largely reduced at the supply voltage of 0.5 V. Furthermore, these results support the need for continued investigations for use of this new technology in achieving the scaling goals dictated by the current ITRS roadmap. Under the assumption that an appropriate choice of gate metal workfunction will further reduce the threshold voltage in these devices, our simulation results predict extremely high drive current with good inherent scalability.

References

1. R. Chau, S. Datta, M. Doczy, B. Doyle, B. Jin, J. Kavalieros, A. Majumdar, M. Metz, and M. Radosavljevic, Benchmarking nanotechnology for high-performance and low-power logic transistor applications, *IEEE Trans. Nanotechnol*, **4**(2), 153-158 (2005).
2. Z. Yu, C. M. Overgaard, R. Droopad, M. Passlack, and J. K. Abrokwah, Growth and physical properties of Ga$_2$O$_3$ thin films on GaAs(001) substrate by molecular-beam epitaxy, *Appl. Phys. Lett.*, **82**(18), 2978-2980 (2003).
3. M. Passlack, J. K. Abrokwah, R. Droopad, Z. Yu, C. Overgaard, S. I. Yi, M. Hale, J. Sexton, and A. C. Kummel, Self-aligned GaAs p-channel enhancement mode MOS heterostructure field-effect transistor, *IEEE Electron Device Lett.*, **23**(9), 508-510 (2002).
4. M. Passlack, Methodology for development of high-κ stacked gate dielectrics on III-V semiconductors, in *Materials Fundamentals of Gate Dielectrics*, A. A. Demkov and A. Navrotsky, Eds., Dortrecht, The Netherlands: Springer, 403-467 (2005).
5. M. Passlack, M. Heyns, and I. Thayne, III-Vs and Ge look to help CMOS, *Compound Semiconductor,* **14**(4), 21-24 (2008).
6. M. Passlack, O, Hartin, M. Ray, and N. Medendorp, Enhancement mode metal-oxide-semiconductor-field effect transistor, U.S. Patent 6 963 090, Nov. 8, 2005.

7. M. Passlack, K. Rajagopalan, J. Abrokwah, and R. Droopad, Implant-free, high mobility flatband MOSFET: Principles of operation, *IEEE Trans. Electron Dev.*, **53**(10), 2454-2459 (2006).

8. K. Kalna, M. Boriçi, L. Yang, and A. Asenov, Monte Carlo simulations of III-V MOSFETs, *Semicond. Sci. Technol.*, **19**(4), S202-S205 (2004).

9. K. Kalna, S. Roy, A. Asenov, K. Elgaid, and I. Thayne, Scaling of pseudomorphic high electron mobility transistors to decanano dimensions, *Solid-State Electron.*, **46**(5), 631-638 (2002).

10. D. A. J. Moran, K. Kalna, E. Boyd, F. McEwan, H. McLelland, L. L. Zhuang, C. R. Stanley, A. Asenov, and I. Thayne, Self-aligned 0.12 μm T-gate In$_{.53}$Ga$_{.47}$As/In$_{.52}$Al$_{.48}$As HEMT technology utilizing a non-annealed ohmic contact strategy, *in Proc. ESSDERC 2003*, J. Franca and P. Freitas, Eds., Estoril, Portugal, 315-318 (2003).

11. K. Kalna, K. Elgaid, I. Thayne, and A. Asenov, Modelling of InP HEMTs with high Indium content channels, *in Proc. 17th Indium Phosphide and Related Materials Conf.*, A. Marsh and I. Thayne, Eds., 61-65 (2005).

12. D. K. Ferry, Effective potentials and the onset of quantization in ultrasmall MOSFETs, *Superlatt. Microstruct.*, **28**(5-6), 419-423 (2000).

13. K. Kalna, L. Yang, and A. Asenov, Monte Carlo simulations of sub-100 nm InGaAs MOSFETs for digital applications, *in Proc. ESSDERC 2005*, G. Ghibaudo T. Skotnicki, S. Cristoloveneanu, and M. Brillouët, Eds., 169-172 (2005).

14. I. Knezevic, D. Z. Vasileska, and D. K. Ferry, Impact of strong quantum confinement on the performance of a highly asymmetric device structure: Monte Carlo particle-based simulation of a focused-ion-beam MOSFET, *IEEE Trans. Electron Devices*, **49**(6), 1019-1026 (2002).

15. M. Passlack, Development methodology for high-κ gate dielectrics on III-V semiconductors: Gd$_x$Ga$_{0.4-x}$O$_{0.6}$/Ga$_2$O$_3$ dielectric stacks on GaAs. *J. Vac. Sci. Technol. B*, **23**(4), 1773-1781 (2005).

International Journal of High Speed Electronics and Systems
Vol. 19, No. 1 (2009) 101–106
© World Scientific Publishing Company

World Scientific
www.worldscientific.com

THE FIRST 70NM 6-INCH GaAs PHEMT MMIC PROCESS

H. KARIMY, L. GUNTER, D. DUGAS, P.C. CHAO, W. KONG, S. YANG, P. SEEKELL,
K.H.G. DUH, J. LOMBARDI, L. MT PLEASANT
BAE Systems, Nashua, NH (USA) hameed.f.karimy@baesystems.com

BAE Systems has developed a high power, high yield 70nm 6" 2-mil PHEMT MMIC process for frequencies up to 100GHz. Utilizing T-gate technology and 2-mil substrates, we have created a millimeter wave technology that produces excellent performance from Ka-band through W-bands. The device DC and RF characteristics have excellent uniformity across the wafer. In this paper, we report the 70nm device fabrication on 6-inch wafers and compare the DC and RF characteristics with its mature 0.1μm counterpart.

Keywords: W-band; PHEMT; GaAs; Power Amplifier; Millimeter-Wave; MMIC

1. Introduction

In order to be competitive, semiconductor technology paths are driven by both cost reduction and increased performance. Six-inch (150mm) GaAs wafer technology has generated significant interest, since it offers very high performance PHEMT MMICs at low cost compare to 3" and 4" wafers (see Table 1) for commercial and military applications [1,2]. Due to the challenges in fine line lithography, existing 6-inch PHEMT technology is limited to 0.1μm gate length, and therefore usable only up to ~60GHz. To improve device and MMIC performance beyond 60GHz, the device gate length on a 6-inch wafer needs to be reduced to <0.1μm, creating for high yield, very short gate fabrication and scaling in both material epitaxial and device structures. In this paper, we discuss the successful development of the first 70nm PHEMT devices and MMICs on 6-inch GaAs substrates which achieves high performance up to 100GHz

Table 1. Cost advantage of 6-inch GaAs process

	3"	4"	6"
Usable wafer*	0.6	1	2.4
Process Cost	1.0	1	1.2
Epi Cost	0.8	1	1.8
Chip Out Per Lot	0.6	1	2.4
Chip Unit Cost**	1.6	1	0.6

2. Power PHEMT Technology and Process

The 70nm 6-inch single-recess (SR) PHEMT MMIC fabrication utilizes mostly cassette-to-cassette equipment in the foundry. The highly automatic wafer processing reduces human wafer handling and improves both uniformity and visual yield. The epitaxial layers are grown via MBE and have been optimized to reduce the short channel effect for the 70nm device. They are based on a double-heterojunction epi structure with doping on both sides of the InGaAs channel to provide a higher sheet charge density for higher full channel current and output power. Ohmic contacts are formed using a Ge-Au based process with two metal layers for interconnects. PHEMTs are passivated using PECVD SiN followed by creation of MIM capacitors with unit capacitance of 400pF/mm^2. The 70nm T-gates are realized through a unique two-step e-beam lithography process[3], a fully selective gate recess, and Ti/Pt/Au metallization. Particular attention was paid during the gate recess process to minimize the surface trapping effect. All of the wafers were tested after passivation for pulsed I-V to validate the minimization of surface trapping. Figure 1 shows the SEM cross-section of a 70nm PHEMT.

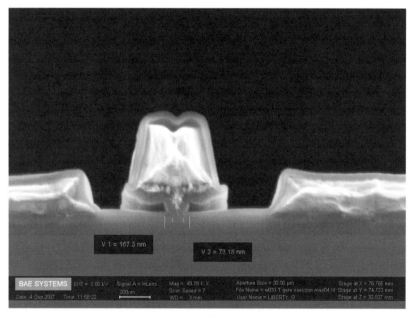

Fig. 1. Cross sectional view of a 70nm PHEMT gate on a 6-inch GaAs wafer

As shown in Table 2, the processed wafers have excellent gate length uniformity within the wafer and from wafer-to-wafer. It can be noted that the gate length in resist is less than 70nm. This is to account for metal skirting during gate deposition to produce a final footprint of 70nm, as shown in Figure 1.

Table 2. Uniformity of resist opening for 70nm gate (target for resist opening before gate recess is 60nm)

Wafer #	36	5	49
Site 1	63nm	61nm	64nm
Site 2	64	56	66
Site 3	67	66	62
Site 4	57	57	55
Site 5	64	58	57
Site 6	56	59	58
Site 7	61	61	57
Site 8	54	65	61
Site 9	56	58	58
Site 10	68	63	62
Site 11	61	59	59
Site 12	59	67	55
Site 13	59	58	56
Site 14	60	59	62
Site 15	58	61	62
AVG	60.47	60.53	59.60
Max	68	67	66
Min	57	56	55
STD	4.10	3.36	3.38

Wafers were then backside processed using a 2-mil substrate process with optimized 10µm wide source vias to provide direct grounding for reduced source inductance and improved thermal resistance – critical for very high frequency, high power applications. A new technique has been developed to handle the fragile 2-mil thin 6-inch wafers after demount for RF on-wafer testing.

3. Device Characteristics

As shown in Figure 2, the 70nm SR PHEMT exhibited an extrinsic transconductance of 710mS/mm with $I_{ds,max}$ of 700mA/mm.

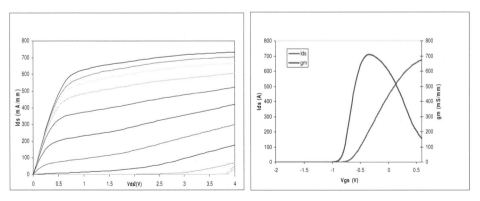

Fig. 2. DC characteristic of a 70nm gate PHEMT with VgsStart: 0.8v, VgsStop: -1.8 v, VgsStep: -0.2v, Vds=2.5v and device Size: 2 x 75µm

Figures 3 and 4 compare the device pulsed I-V characteristics (which are critical parameters for high power performance). As can be seen from these figures, other than a minor knee voltage walkout, the pulsed I-V characteristics of the 70nm does not exhibit current collapse – similar to the 0.1μm PHEMT. Off-state gate-drain breakdown of 11V and three-terminal on-state breakdown of 4.5V were measured with no observation of the short channel effect. Small signal RF measurement indicates that due to the shorter gate length, the 70nm device performs better at higher frequencies – with F_{max} of 200GHz, compare to 150GHz for the 0.1μm device.

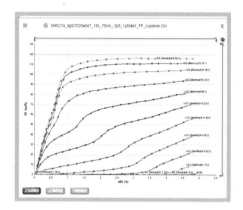
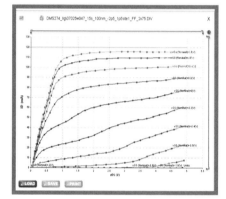

Fig. 3. Pulsed I-V characteristics of a 70nm PHEMT   Fig. 4. Pulsed I-V characteristics of a 100nm PHEMT

The comparison in DC and RF characteristics are summarized in Table 3.

Table 3. DC Characteristics: 100nm vs. 70nm gate

Parameter	Unit	100nm	70nm
I_{max}	mA/mm	615	700
g_m	mS/mm	570	710
V_{bd}	V	11	11
F_{max}	GHz	150	200
F_t	GHz	90	100

4. MMIC Results

Several 70nm PHEMT device process runs were performed on wafers with slight variations in the material structure to generate the initial small-signal and non-linear models. These early models allowed successful first-pass designs to be inserted on a 70nm MMIC fabrication run. Our expectation was that these first-pass MMICs would function well, but not at optimum conditions. Many FET structures and other diagnostics were placed on the mask to allow for further and MMIC design optimization on a second iteration.

Measured results for these initial MMICs are quite encouraging and on-wafer results for two MMIC circuits are shown in Figures 5 and 6. Each circuit is a 4-stage MMIC with the periphery of each stage totaling 160µm. The circuit is biased at a drain voltage of 3.5V and a current of 128mA (200mA/mm). The MMIC in Figure 5 has approximately 20 dB of gain from 60 GHz to 90GHz. The MMIC in Figure 6, while mistuned, shows approximately 16 dB of gain at 100GHz which is impressive for a PHEMT based circuit. Once FET data is collected, and new models generated, the circuits can be optimized for a second fabrication cycle. Early indications are that the MMIC gain shown in Figures 5 and 6 will be improved with a second design pass. The 70nm PHEMT process described in this paper will allow MMICs to operate up to 110 GHz. These circuits were processed on a 6-inch wafer thereby providing a low cost solution to W-band MMICs. The yield for two wafers that were RF tested were respectively 77% and 83%. The average yield of 80% is an excellent result for the first pass on these MMICS.

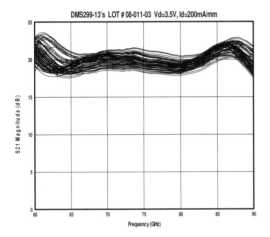
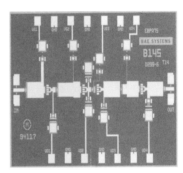

Fig. 5. The MMIC has approximately 20 dB of gain from 60 GHz to 90GHz, output stage= 0.640 mm

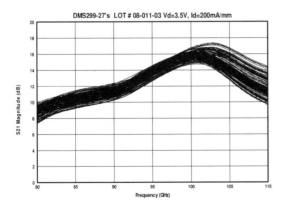
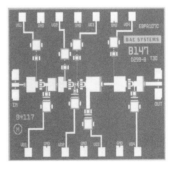

Fig. 6. The MMIC shows approximately 16 dB of gain at 100GHz, output stage=0.640 mm

5. Conclusion

With an advanced gate lithography technique and a proper material and device scaling, we have demonstrated for the first time a high performance 70nm 6-inch GaAs PHEMT MMIC process with high yield. Using this technology, we have produced PHEMT MMICs with excellent gain output power operating at mm-wave frequencies. The demonstration of the 70nm power PHEMT MMIC technology on 2-mil thick 6-inch GaAs substrates is a significant milestone since it makes possible the availability of very high performance PHEMT MMIC for up to 100GHz applications at low cost.

Acknowledgment

We would like to thank Matt Morgan and Eric Bryerton with the National Radio Astronomy Observatory for their contribution. "The National Radio Astronomy Observatory is facility of the National Science Foundation operated under cooperation agreement by Associated Universities, Inc."

References

1. M. Chertouk, et. al., "Manufacturable 0.15μm PHEMT Process for High Volume and Low Cost on 6" GaAs Substrates," in Proc. GaAs ManTech Conf., May 2002.
2. L. Gunter, et.al, "The First 0.1μm 6" GaAs PHEMT MMIC Process," in Proc. GaAs ManTech Conf., Apr. 2006.
3. D. Xu., et. al, "A reproducible, high yield method for fabricating ultra-short T-gates on HFETs," patent pending

International Journal of High Speed Electronics and Systems
Vol. 19, No. 1 (2009) 107–112
© World Scientific Publishing Company

HIGH-PERFORMANCE 50-NM METAMORPHIC HIGH ELECTRON-MOBILITY TRANSISTORS WITH HIGH BREAKDOWN VOLTAGES

DONG XU, WENDELL M.T. KONG, XIAOPING YANG, P. SEEKELL, L. MOHNKERN, H. KARIMY, K.H.G. DUH, P.M. SMITH, and P.C. CHAO

Microelectronics Technology & Products, Electronics and Integrated Solutions,
BAE Systems, Nashua, NH 03060

This paper reports a successful improvement of the low breakdown voltages in short gate-length metamorphic high electron-mobility transistors. The technical approach includes both the optimization of the epitaxial layer design and the selection of the proper gate recess scheme. By employing a novel epitaxial design (including a high indium composite channel and the double-sided doping) and an asymmetric gate recess, both the off-state and on-state breakdown voltages have been improved for 50-nm high-performance metamorphic high electron-mobility transistors. The results reported herein demonstrate that these devices are excellent candidates for ultra-high-frequency power applications.

Keywords: High electron-mobility transistor; breakdown voltage

1. Introduction

InP-based high electron-mobility transistor (HEMT) and GaAs-based metamorphic electron-mobility transistor (MHEMT) with indium-rich channel designs are well known for their outstanding low noise [1] and high-gain performance [2], as a result of the superior transport properties associated with the InAlAs/InGaAs heterostructures. However, this materials system also limits the power performance of the HEMT because of its low breakdown voltage and enhanced impact ionization, which is associated to the low band gap. This problem becomes increasingly critical as the gate length is reduced to below 100-nm while the channel indium content is increased to boost gain for ultra-high-frequency operation. So far, most of the state-of-the-art results of high-indium InP HEMTs and high-indium MHEMTs are limited to the high transconductance and high cut-off current gain frequency for high-speed application. For example, Yamashita et al. succeeded in fabricating a 25-nm pseudomorphic HEMT with 562 GHz current gain cut-off frequency and gm of 1.23 S/mm [3]. However, the gate-drain breakdown voltage was less than 1 V (defined at 1 mA/mm gate current) and the output characteristics measured only to a maximum drain bias of 0.8V, indicating a low on-state breakdown voltage. It was also reported recently that InP HEMTs with a 35-nm gate length had achieved high gain of 4.4 dB at 308 GHz, while the off-state breakdown voltage had improved to 2.5V (defined at 0.25 mA/mm gate current) and the on-state breakdown voltage was about 1.8V (with gate current of 0.25 mA/mm and drain current of 290 mA/mm) [4].

Efforts have been made to improve the breakdown voltage performance of InP HEMTs in the past several years. The technical approaches include the employment of the novel channel designs and the use of a wider recess groove. However, the use of the low indium material layer [5] or other wide bandgap channel materials like InP or InAsP layers [6] as the sub-channel normally degrades the overall transport properties of the channel, lowering the device gain performance. We selected high-indium InGaAs layers as channel materials and placed Si spike doping below and above the channel. We also chose a wider recess groove to help increase the breakdown voltages, but worked to minimize its negative impact on small gate length devices. In this paper, we report that the breakdown voltage of high-performance 50-nm MHEMTs was significantly improved with an epitaxial design optimization and the implementation of an asymmetric gate recess technology.

2. Device Fabrication

Molecular beam epitaxy (MBE) was used to grow the MHEMT structures for this work. Several epitaxial designs were grown, including a single-sided doped structure with an 80% InGaAs channel (structure A) and a double-sided doped structure with an high-indium composite channel (structure B). All of the epitaxial HEMT structures were grown on GaAs substrates after a thick buffer layer was initially grown. The typical room temperature sheet carrier density was around 3.3×10^{12} and 3.8×10^{12} cm^{-2} for structure A and structure B, respectively. The corresponding room temperature electron mobility was around 11500 and 12000 cm^2/Vs, respectively. It is worth noting that structure B had both higher sheet carrier density and electron mobility in comparison to structure A.

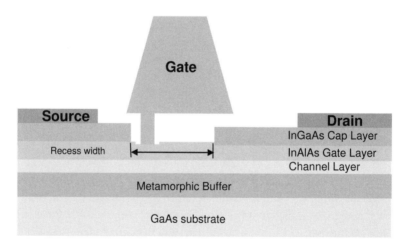

Fig. 1. Asymmetrically recessed 50-nm MHEMT fabricated on active layers grown on metamorphic buffer. The passivation layer is not shown.

The device fabrication process included conventional steps such as mesa isolation, ohmic metal, gate definition, passivation, and interconnect metal deposition. We employed an asymmetric gate recess scheme that created a wider recess on the drain side of the gate and a narrower recess on the source side of the gate to boost the breakdown voltage, minimize source resistance, and reduce output conductance degradation caused by the 50-nm gate length. The asymmetric gate recess was implemented with a separate electron beam lithography step. After the removal of the highly-doped cap layer, a two-step exposure and developing process based on a tri-layer electron beam resist scheme was performed, resulting in an excellent gate metal liftoff. Figure 1 shows a schematic of the MHEMT with an asymmetrically recessed Γ-gate. The recess width was varied with the electron-beam lithography definition, and the gate length was nominally 50 nm.

3. Results and Discussions

The two-terminal off-state breakdown voltage, BV_{off}, defined as the gate-drain voltage at which a gate current of 1 mA/mm is reached with the source electrode floating, for devices fabricated on both structures is summarized in Figure 2. On structure A, the device with a 300-nm recess width had a BV_{off} around 4 V higher compared to a device with a 150-nm recess width. A similar trend was also observed for devices fabricated on structure B. This is because the off-state breakdown is basically governed by the extension of the depletion area between the gate and the drain, which is defined by the recess groove on the drain side. It is also worth noting that the device on structure B showed a slightly higher BV_{off} compared to structure A with the same recess width.

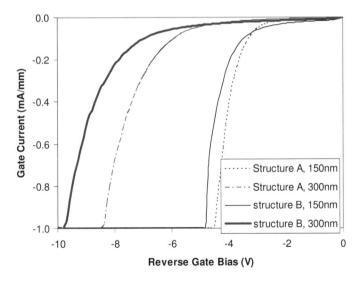

Fig. 2. Reversed diode characteristics for 50-nm MHEMTs fabricated on structures A and B with recess groove widths of 150 nm and 300 nm.

Measurements of the on-state breakdown, BV_{on}, defined as the drain bias at which the gate current reaches 1 mA/mm at the gate bias for peak transconductance, are shown in Figure 3. The devices fabricated on structure A appeared to be insensitive to the variation of the recess width. However, a substantially higher BV_{on} of around 3.5 V was achieved on structure B for devices with a 300-nm wide recess groove, which should be attributed to its composite channel design that reduces the impact ionization in the channel layers [7]. High BV_{on} of the InGaAs channel HEMTs with ultra-short gate lengths could be achieved with the optimization of the epitaxial design and the fabrication technology.

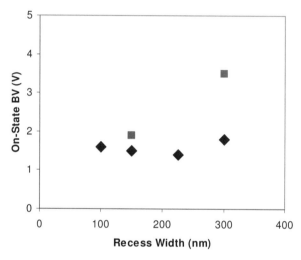

Fig. 3. On-state breakdown voltage for 50-nm MHEMTs fabricated on structures A (diamonds) and B (squares) with different recess groove widths.

Figure 4 shows the excellent output characteristics of the MHEMT device with enhanced breakdown voltages using the optimized epitaxial design (structure B) and the optimized 300-nm asymmetrical gate recess groove. Generally speaking, the larger recess would lower the transconductance and the drain current, but proper processing consideration can substantially alleviate this adverse influence, allowing us focus more on the improvement of high breakdown performance. The high-performance DC characteristics of this device can be summarized as follows: greater than 9V BV_{off}, around 3.5V BV_{on}, approximately 800 mA/mm maximum drain current, 1.9 S/mm peak transconductance at a drain bias ranging from 1-2 V, and a low output conductance of 25 mS/mm at a drain bias of 2 V.

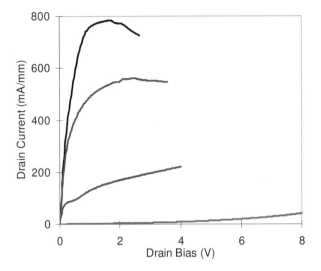

Fig. 4. Output characteristics for 50-nm MHEMTs fabricated on structure B with recess groove width of 300 nm. The gate bias for the top curve is 0.4 V, and the gate bias step is -0.2 V.

The drain current and peak transconductance of our optimized MHEMT devices are at least comparable to those of ultra-short InP-based HEMTs optimized for high gm and high-speed performance [3-5], but the breakdown voltages we report here are substantially higher. Actually, the breakdown voltages are even better than the 0.07-μm InP-based power HEMTs utilizing wider bandgap channel materials like InP or InAsP, which, however, have much lower drain current and transconductance [6], imposing significant restrictions to their possible applications at high frequency. It is also worth noting that the voltage gain (the ratio of transconductance and output conductance) of this MHEMT has a record high value of about 80 at a drain bias of 2 V, comparing favorably with the value of 5.2 determined for the 35-nm gate InP HEMT [4] and value of 10.5 of the asymmetric 50-nm MHEMT at a drain bias of 1 V [2], which indicates very high gain performance.

4. Summary

This work demonstrates that excellent breakdown voltages are achievable on 50-nm MHEMTs without compromising the state-of-the-art device characteristics, including drain current, peak transconductance and voltage gain. The simultaneous realization of high performance and high breakdown makes these devices excellent candidates for ultra-high-frequency power applications in the future.

5. Acknowledgments

The authors are also grateful to R. Carnevale and R. Isaak for layout work and K. Beech, D. Gallagher, M. Gerlach, J. Pare, J. Brahmbhatt, J. Hulse and B. Golja for their processing support.

References

1. M. Schlechtweg, A. Tessmann, A. Leuther, C. Schwoerer, M. Lang, U. Nowotny and O. Kappeler, Integrated circuits based on 300 GHz f_T metamorphic HEMT technology for millimeter-wave and mixed-signal applications, *Proc. 11ᵗʰ Int. GaAs Symp.*, Munich, Germany, 2003.
2. Dong Xu, Wendell M.T. Kong, Xiaoping Yang, P.M. Smith, D. Dugas, P.C. Chao, G. Cueva, L. Mohnkern, P. Seekell, L. Mt. Pleasant, B. Schmanski, K.H.G. Duh, H. Karimy, A. Immorlica and J.J. Komiak, Asymmetrically recessed 50-nm gate-length metamorphic high electron-mobility transistor with enhanced gain performance, *IEEE Electron Device Lett.*, **29**(1), 4-8(2008).
3. Y. Yamashita, A. Endoh, K. Shinohara, K. Hikosaka, T. Matsui, S. Hiyamizu and T. Mimura, Pseudomorphic $In_{0.52}Al_{0.48}As/In_{0.7}Ga_{0.3}As$ HEMTs with an ultrahigh f_T of 562 GHz, *IEEE Electron Device Lett.* **23**(10), 573-575(2002).
4. X.B. Mei, W. Yoshida, W.R. Deal, P.H. Liu, J. Lee, J. Uyeda, L. Dang, J. Wang, W. Liu, D. Li, M. Barsky, Y.M. Kim, M. Lange, T.P. Chin, V. Radisic, T. Gaier, A. Fung, L. Samoska and R. Lai, 35-nm InP HEMT SMMIC amplifier with 4.4dB gain at 308 GHz, *IEEE Elec. Dev. Letters,* **28**(6), 470-472(2007).
5. K. Elgaid, H. McLelland, M. Holland, D.A.J. Moran, C.R. Stanley and I.G. Thayne, 50-nm T-gate metamorphic GaAs HEMTs with f_T of 440 GHz and noise figure of 0.7 dB at 26 GHz, *IEEE Electron Device Lett.*, **26**(11), 784-786(2005).
6. Farid Medjdoub, Mohammed Zaknoune, Xavier Wallart, Chrisophe Gaquiere, Francois Dessenne, Jean-Luc Thobel and Didier Theron, InP HEMT downscaling for power application at W band, *IEEE Trans. Electron Device* **52**(10), 2136-2143(2005).
7. Gaudenzio Meneghesso, Andrea Neviani, Rene Oesterholt, Mehran Matloubian, Takyiu Liu, Julia J. Brown, Claudio Canali, and Enrico Zanoni, On-state and off-state breakdown in GaInAs/InP composite-channel HEMT's with variable GaInAs channel thickness, *IEEE Trans. Electron Device* **46**(1), 2-9(1999).

International Journal of High Speed Electronics and Systems
Vol. 19, No. 1 (2009) 113–119
© World Scientific Publishing Company

MBE GROWTH AND CHARACTERIZATION OF Mg-DOPED III-NITRIDES ON SAPPHIRE

X. CHEN, K. D. MATTHEWS, D. HAO, W. J. SCHAFF, L. F. EASTMAN

School of Electrical and Computer Engineering, Cornell University
Ithaca, NY 14853, USA
xc32@cornell.edu

W. WALUKIEWICZ, J. W. AGER, and K. M. YU

Lawrence Berkeley National Laboratory, Berkeley, CA 94720, USA

Plasma-assisted molecular beam epitaxial growth of Mg-doped GaN and InGaN on a sapphire substrate is investigated in this study. Electrical characteristics of p-type GaN strongly depend on the flux of Mg acceptors and the growth temperature. Only the intermediate range of Mg fluxes (beam equivalent pressures near 1×10^{-9}T) produce p-type GaN with good electrical properties, and a maximum hole concentration of 3.5×10^{18} cm^{-3} is obtained with a Hall mobility of 2.1 cm^2/V·s. Due to the strong surface accumulation of electrons, Hall measurements do not indicate p-type polarity for In fraction beyond 11%. In contrast, hot probe measurements show that p-polarity can be measured for the entire range of Mg-doped In mole fractions. Electroluminescence also indicates p-polarity for Ga-rich mole fractions. In$_x$Ga$_{1-x}$N p-n homojunctions are fabricated and tested. All GaN devices show low series resistance (0.03 ohm-cm^2) and insignificant parasitic leakage. IV curves of all three InGaN homojunctions show rectifying characteristics under dark conditions and photo-response under outdoor sunlight, indicating the existence of holes in InGaN with up to 40% In content.

Keywords: Molecular beam epitaxy; InGaN; Mg-doped

1. Introduction

The direct band gap of III-nitrides extends from InN (0.6 eV, near-IR), to GaN (3.4eV, mid-UV) [1, 2] and finally to AlN (6.2 eV, deep-UV) [3]. Due to such a vast range of band gaps, III-nitride alloys have been intensively studied for their applications for opto-electrical devices such as LEDs and laser diodes, and have undergone remarkable development [4]. Recently, InGaN and InAlN alloys have attracted attention for their potential in solar cell devices [5,6]. In the past, limitations to their application have been the inability to grow high quality epitaxial films as well as to achieve p-type doping. It is necessary to investigate the effects of p-type doping of InGaN.

In 1989, Amano et al. discovered the possibility of achieving p-type doping conductivity in Mg-doped GaN layers grown by metal organic vapor phase epitaxy (MOVPE) [7]. A thermal annealing under N$_2$ atmosphere has been shown to be very efficient to activate the Mg dopants by eliminating Mg-H complexes [7]. However, there is no need of thermal annealing to get p-type conductivity grown by molecular beam epitaxy (MBE) when using an Mg source. State of the art Mg-doped GaN layers exhibit a hole concentration and a hole mobility of 1.4×10^{18} cm^{-3} and 7.5 cm^2 V s for RF-plasma MBE, and 8×10^{17} cm^{-3} and 26 cm^2/V·s for NH$_3$ MBE, respectively [9,10]. In the past, p-doping in In-rich InGaN has proven extremely difficult, with success reported only recently [6]. Also the electrical properties of InGaN are challenging to interpret due to a strong surface accumulation of electrons existing throughout most of the Indium (In) composition range [11]. The most common technique to determine p-type polarity is Hall measurement. The unambiguous p-type polarity by Hall measurement is only seen in Mg-doped InGaN for In content smaller than 11%. Beyond this point, the polarity determined by the magnetic field is either inconsistent or shows n-type polarity.

2. Experimental

2.1. *MBE growth*

GaN and In$_x$Ga$_{1-x}$N are grown on 2-inch c-plane sapphire substrates via MBE with an EPI Unibulb RF plasma source. Sapphire substrates are metalized on the backside with approximately 1 μm thick sputtered tungsten. The wafers are loaded with no surface treatment into a preparation chamber for 1 hour baking at 300 °C in UHV. Following a 30-min RF plasma exposure at 200 °C in the main chamber, the substrate temperature is ramped to 800 °C and AlN and GaN buffer layers are grown to a thickness of 20 nm and 1 μm, respectively. Plasma power is 300 - 400 W and nitrogen flow rate is 0.8 - 1.0 sccm. Then Mg-doped GaN or InGaN film is deposited at relatively lower substrate temperature (between 500-600°C) for approximately two hours to achieve 600 nm thick layers. Substrate temperatures are measured by an optical pyrometer for high temperature growth and thermocouple for low temperature growth.

P-type doping conditions have been optimized during InGaN growth. The Mg source temperature is selected to be in a range that results in the maximum resistivity when In mole fraction beyond 11% is utilized. If the Mg density is too low, background n-type conductivity is high. Too much Mg also results in high net n-type conductivity for In mole fractions beyond 11%, presumably due to donor-like defects. The goal is to compensate, or exceed, the background electron density of $10^{17} - 10^{18}$ cm^{-3} at the InN composition end point. The beam flux pressure for an optimum Mg flux is in the range of $1\text{-}5 \times 10^{-9}$ Torr.

2.2. *Material characterizations*

The identification of the dominant carrier type in the film is performed by hot probe, electroluminescence, and Hall measurements with Van der Pauw geometry using In ohmic contacts. Quantitative mobility spectrum analysis (QMSA) of variable field Hall measurements shows that In contacts form suitable Ohmic contacts with the underlying Mg-doped layers despite the presence of surface electrons [12]. Our hot probe setup has been calibrated to test the polarity of known Si and III-V materials and is proven as a reliable method to evaluate the electrical properties of InGaN:Mg [11].

Electroluminescence is also observed by applying needle probes to p-type material and injecting minority carriers.

2.3. *P-i-n homojunction*

To further evaluate the existence of p-type conductivity by Mg doping, a series of p-i-n homojunction structures made of $In_xGa_{1-x}N$ (x = 0, 0.2, 0.3, 0.4) are grown on sapphire using the optimum growth condition. Mg and Si sources are used to produce p-type and n-type carriers, respectively. Following the growth, devices ranging from 1 mm x 1 mm to 4 mm x 4 mm are fabricated via standard photolithography. The metallization consists of 20 nm Pt/120 nm Au for p contacts and 150 nm Ti/900 nm Al/400 nm Mo/550 nm Au for n contacts.

3. Results and discussions

It is found that electrical characteristics of p-type GaN strongly depend on the Mg flux and the growth condition. Only a narrow range of Mg fluxes (480-530°C) produces p-type GaN with good electrical properties. If the Mg flux is too low, the alloy is effectively undoped and exhibits insulating behavior. However, doping with an extremely high flux of Mg results in n-type polarity in the film, most likely due to the bombardment of Mg atoms and creation of point defects, which act as donor states and compensate the holes in the film. With the same Mg flux, a Ga-rich growth condition results in much higher hole density than N-rich growth condition. In addition, 3x3 or 6x6 reconstruction Rheed patterns are observed when Mg-doped GaN is cooled below 400 °C. Fig. 1 summarizes the hall mobility vs. hole concentration of Mg-doped GaN grown on sapphire. Maximum hole concentration of 3.5×10^{18} cm^{-3} is obtained with a Hall mobility of 2.1 cm^2/V s.

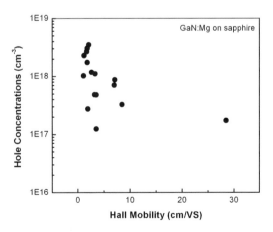

Fig. 1. Hall mobility vs. hole concentration of Mg-doped GaN grown on sapphire

Mg-doped $In_xGa_{1-x}N$ films of composition ranging from x = 0 to x = 0.88 are grown successfully and with high quality as indicated by atomic force microscopy (AFM) and XRD. Fig. 2 shows the SIMS profile of Mg-doped $In_{0.24}Ga_{0.76}N$. The incorporated Mg concentration is as high as 2 x 10^{20} cm^{-3} and Mg doping is shown to be fairly uniform. Hall measurement indicates a hole concentration of $7.7x10^{17}$ cm^{-3} in Mg-doped $In_{0.04}Ga_{0.96}N$. However, when the In mole fraction becomes larger than 0.11, the samples measured by Hall effect exhibit strong n-type polarity, whereas p-type polarity is confirmed by the hot probe measurement for all of the Mg-doped InGaN samples.

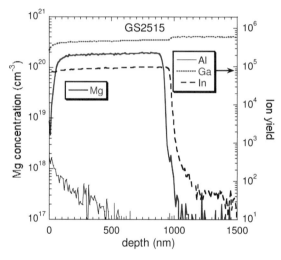

Fig. 2. SIMS profile of Mg-doped $In_{0.24}Ga_{0.76}N$ grown on sapphire

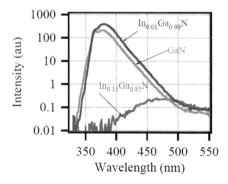

Fig. 3. Electroluminescence from Mg-doped InGaN at a bias of approximately 4V

Figure 3 indicates the electroluminescence from Mg-doped InGaN obtained at a bias of approximately 4V. The bias is applied across a pair of tungsten needle probes located 500 μm apart, and the light is collected from the back of the sapphire substrate. The forward bias voltage is just slightly above band gap energy, but not large enough for any avalanche processes to occur. It is seen that the electroluminescence intensity becomes weaker with increasing mole fraction. Furthermore, light emission is observed when a bias is applied on Mg-doped InGaN across the tungsten probes (pulsed at 1microsecond, 0.1-1% duty cycle to avoid heating), but no light occurs when n-type or undoped InGaN is similarly biased. A possible explanation for these observations is that the holes in the p-type layer recombine with the minority carriers (electrons) injected from the leaky reverse biased Schottky to produce band gap (or band to impurity) electroluminescence. Despite the lack of clear understanding, the consistent observation of electroluminescence only when the material conductivity is p-type by hot-probe provides another method to evaluate the polarity besides the Hall analysis.

Four p-i-n homojunctions made of $In_xGa_{1-x}N$ (x = 0, 0.2, 0.3, 0.4) are fabricated and tested. I-V measurements are taken under dark conditions, outdoor sun light, as well as concentrated sun light. Figure 4 shows the I-V curve of the 1 mm x 1 mm all-GaN device under outdoor sun light. The n and p contact resistivity of this device is measured at 2.0 x 10^{-3} ohm-cm^2 and 3 x 10^{-4} ohm-cm^2, respectively. Series resistance is very low (0.03 Ω-cm^2) and the parasitic leakage is insignificant. Under 100X solar concentration, GaN cells exhibit a fill factor of 60% and a lower limit of conversion efficiency of 0.8%. The efficiency calculation does not account for the angle of the sun, reflection and transmission from the focus lens. No anti- reflection coating has been applied.

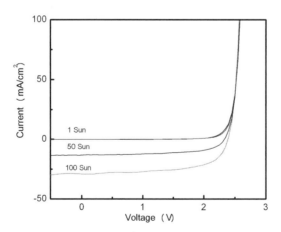

Fig. 4. I-V curves of 1x1 mm all-GaN device in sun light, 1x 50x and 100x concentration.

IV curves of all three InGaN homojunctions show rectifying characteristics under dark condition and photo-responses under outdoor sun light, indicating the existence of holes in InGaN with up to 40% In content. It is well above the 11% In mark which can be detected by Hall measurement. The electroluminescence peak overlaps with the absorption edge, which indicates a relative lack of localized states and phase separation. Absorption measurements are performed on the InGaN cells. From Fig. 5, the absorption edge of $In_{0.24}Ga_{0.76}N$ cell is 2.29 eV. Electroluminescence was also measured by applying a 300 mA forward bias. It is observed that the electroluminescence peak of $In_{0.24}Ga_{0.76}N$ cell overlaps with the absorption edge of the device, which indicates relative lack of localized states.

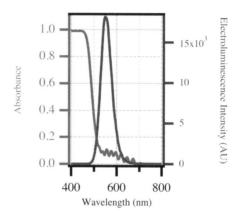

Fig. 5. Absorption and electroluminescence for $In_{0.24}Ga_{0.76}N$ solar cell

4. Conclusions

Electrical characteristics of p-type GaN depend on the flux of Mg acceptors and the MBE growth temperature. Only a narrow range of Mg fluxes produce p-type GaN. A maximum hole concentration of 3.5×10^{18} cm^{-3} is obtained with a Hall mobility of 2.1 $cm^2/V·s$. Due to the strong surface accumulation of electrons, Hall measurements do not indicate p-type polarity for In fraction beyond 11%. In contrast, hot probe measurement shows that p-polarity can be measured for the entire range of Mg-doped In mole fractions. Electroluminescence also indicates p-polarity for Ga-rich mole fractions. The rectified characteristics of IV curves from $In_xGa_{1-x}N$ p-n homojunctions further prove the existence of holes in InGaN with up to 40% In fraction.

5. Acknowledgments

This work was supported by AFOSR through the MURI at Georgia Tech and Phase II SBIR with Structured Materials Inc (SMI) and Rosestreet Energy Labs.

References

1. J. Wu, W. Walukiewicz, K.M. Yu, J.W. Ager III, E.E. Haller, H. Lu, W.J. Schaff, Y. Saito, and Y. Nanishi, "Unusual properties of the fundamental band gap of InN," Appl. Phys. Lett. **80** (21), 3967-3769 (2002).
2. J. Wu, W. Walukiewicz, K. M. Yu, J. W. Ager, S. X. Li, E. E. Haller, H. Lu, and W. J. Schaff, "Universal bandgap bowing in group-III nitride alloys," Sol. St. Comm. **127** (6), 411 (2003).

3. E. Silveira, J. A. Freitas, O. J. Glembocki, G. A. Slack, and L. J. Schowalter, "Excitonic structure of bulk AlN from optical reflectivity and cathodoluminescence measurements," Phys Rev B **71** (4), (2005).

4. S. Nakamura, S. Pearton and G. Fasol, "The blue laser diode: The complete story (2nd and extended edition)" (Springer-Verlag Berlin Heidelberg New York).

5. J. Wu, W. Walukiewicz, K. M. Yu, W. Shan, J. W. Ager III, E. E. Haller, H. Lu, and W.J. Schaff, W. K. Metzger, S. R. Kurtz, and J. F. Geisz, "Superior radiation resistance of In$_{1-x}$Ga$_x$N alloys: a full-solar-spectrum photovoltaic material system," J. Appl. Phys. **94** (10), 6477-6482 (2003).

6. X. Chen, K. Matthews, D. Hao, W. J. Schaff, and L. F. Eastman, "Growth, fabrication and characterization of InGaN/GaN solar cells by MBE," Phys. Stat. Sol. (a) **205** (5), 1103 (2008).

7. H. Amano, M. Kito, K. Hiramatsu, and I. Akasaki, "P-Type conduction in Mg-Doped GaN treated with low-energy electron beam irradiation (LEEBI)," Jpn. J. Appl. Phys. **28** (12), L2112-L2114 (1989).

8. S. Nakamura, T. Mukai, M. Senoh, and N. Iwasa, "Thermal annealing effects on p-type Mg-doped GaN films," Jpn. J. Appl. Phys. **31**, L139-L142 (1992).

9. I. P. Smorchkova, E. Haus, B. Heying, P. Kozodoy, P. Fini, J. P. Ibbetson, S. Keller, S. P. DenBaars, J. S. Speck, and U. K. Mishra, "Mg doping of GaN layers grown by plasma-assisted molecular-beam epitaxy," Appl. Phys. Lett. **76** (6), 718-720 (2000).

10. W. Kim, A. E. Botchkarev, A. Salvador, G. Popovici, H. Tang, and H. Morkoç, "On the incorporation of Mg and the role of oxygen, silicon, and hydrogen in GaN prepared by reactive molecular beam epitaxy," J. Appl. Phys. **82** (1), 219-226 (1997).

11. W. J. Schaff, X. D. Chen, D. Hao, K. Matthews, T. Richards, L. F. Eastman, H. Lu, C. J. H. Cho, and H. Y. Cha, "Electrical properties of InGaN grown by molecular beam epitaxy," Phys. Stat. Sol. (b) **245** (5), 868-872 (2008).

12. P. A. Anderson, C. H. Swartz, D. Carder, R. J. Reeves, S. M. Durbin, S. Chandril and T. H. Myers, "Buried *p*-type layers in Mg-doped InN," Appl. Phys. Lett. **89** (18), 184104 (2006).

International Journal of High Speed Electronics and Systems
Vol. 19, No. 1 (2009) 121–127
© World Scientific Publishing Company

PERFORMANCE OF MOSFETs ON REACTIVE-ION-ETCHED GaN SURFACES

KE TANG, WEIXIAO HUANG[†], T. PAUL CHOW

Center for Integrated Electronic, Rensselaer Polytechnic Institute
Troy, NY 12180, USA
tangk2@rpi.edu

We have fabricated, characterized and compared the performance of lateral enhancement-mode GaN MOSFETs on as-grown and RIE-etched surfaces with 900 and 1000°C gate oxide annealing temperatures. Both subthreshold swing and field effect mobility show 1000°C is the optimal annealing temperature for the PECVD gate oxide in our MOSFET process.

Keywords: GaN; MOSFET; RIE; Subthreshold swing; Field-effect mobility; Annealing

1. Introduction

The wide energy band gap (3.4 eV) and large critical electric field (3 MV/cm) of GaN make itself attractive for high temperature and high-voltage devices. Silicon power devices are reaching their material limits with the development of power devices and power electronics. Wide band gap materials such as SiC and GaN are therefore widely studied as an alternative for power electronics applications. Among GaN-based electronic devices, GaN high electron mobility transistor (HEMT) attracted most of the attention due to the two dimensional electron gas (2DEG) of the AlGaN/GaN heterostucture, which has a better trade-off between on resistance and reverse blocking capability compare to GaN MOSFET. However, GaN MOSFET is superior to GaN HEMT for its low gate leakage and positive threshold voltage for normally-off operation. Therefore, the combination of GaN MOSFET and HEMT will have the advantages of these two, which leads to the idea of GaN hybrid MOS-HEMT transistor [1]. The critical step of making this transistor successfully is RIE etching of the recess gate region and the subsequent damage removal.

Previously we have demonstrated GaN MOS with low interface densities ($\sim 10^{10}$/cm^2-eV) and subsequent high performance long-channel enhancement-mode n-channel MOSFETs on p and n- GaN/sapphire substrates with maximum field-effect mobility of 167 cm^2/V-s, BV up to 940 V and reverse blocking capability [2-4]. More recently, we reported the negative C-V shift, C-V stretch out, higher interface densities, and degraded field-effect mobility on RIE etched GaN MOS capacitors and MOSFETs. We also reported the wet

---

[†] Present Address: Freescale Semiconductor, Tempe, AZ 85284, USA, E-mail:Weixiao.Huang@freescale.com

etch treatment can only partially recover the dry etch introduced damages [5]. In this paper, we will focus on the output IV characteristics, subthreshold swing and field-effect mobility of GaN MOSFETs on dry/wet etched and unetched GaN surfaces.

2. Device Structures and Fabrication

Devices with varying channel lengths were fabricated on 30 nm UID $Al_{0.22}Ga_{0.78}N$ on 3 μm UID GaN on sapphire substrate (MOSFET 1), 0.3 μm $1\times10^{17}cm^{-3}$ n-GaN on 6 μm $1\times10^{16}cm^{-3}$ p-GaN on sapphire substrate (MOSFET 2), and 3 μm UID GaN on sapphire substrate (MOSFET 3). The schematic cross-section view and microphotograph are shown in Figs. 1 and 2 respectively.

ICP etch with depths of 50nm was performed to define gate regions for MOSFET 1 and 350 nm ICP etching were performed for MOSFET 2 and 3 entire surfaces, followed by wet etching to remove damages from dry etching. $POCl_3$ doped polysilicon was used as gate electrode. 900°C anneal for 30 min. in N_2 was performed on 100 nm gate oxide for MOSFET 1 while 1000°C anneal for 30 min. in N_2 was performed on 100 nm gate oxide for MOSFET 2 and 3. In addition, devices utilizing 600 nm field oxide as the gate insulator were also fabricated and inspected in MOSFET 1 and 2. Ti/Al was deposited and annealed at 600°C for 10 min. for n^+ ohmic contacts.

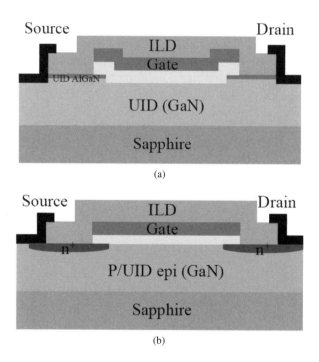

(a)

(b)

Fig. 1. Cross-section view of (a) MOSFET 1, and (b) MOSFET 2, 3

Fig. 2. Microphotograph of GaN MOSFET.

3. Experimental Results

All three MOSFETs discussed in this paper are self enclosed circular structure with drain in the center and is surrounded by gate (channel region), and source is at the edges of the circle. This structure has the benefit of low leakage current compare to the conventional linear structure. For comparison, we characterized the three MOSFETs on the same dimension devices with channel widths of 800 µm and channel lengths of 100 µm in order to avoid the series resistance effect under the purpose of extracting field-effect mobility, and subthreshold slope.

Figure 3 shows the output IV characteristics of all three MOSFETs with drain to source voltage ranged from 0V to 10V, and gate to source voltage steps from 0V to 30V for every 5V. The current level at V_{GS} = 30V is largest for MOSFET 3 which is with as-grown surface, and lowest for MOSFET 1 which is with dry/wet-etched surfaces and gate oxide annealed at 900°C. The on-resistances are 2944 Ω-mm, 1333 Ω-mm, and 970 Ω-mm for MOSFET 1, 2 and 3 respectively.

Figure 4 shows the I_{DS} and G_M vs. V_{GS} curves for all three MOSFETs. From the I_{DS} curves, we found that MOSFET 1 has a soft turn-on characteristics comparing to MOSFET 2 and 3. The maximum I_{DS} and G_M of MOSFET 1 are also much smaller than the other two. They are 38.3 µA, 82.1 µA, 92.2 µA for maximum I_{DS}, and 1.49 µS, 3.12 µS, 3.74 µS for maximum G_M in the order of MOSFET 1, 2 and 3.

Fig. 5 shows the 100 nm gate oxide devices subthreshold swing characteristics of all three MOSFETs. The subthreshold slope of MOSFET 1 (770mV/decade) is higher than those in MOSFET 2 (300mV/decade) and MOSFET 3 (380mV/decade). Similarly, in field oxide devices, the subthreshold slope of MOSFET 1 (4.8V/decade) is higher than that in MOSFET 2 (1.5V/decade).

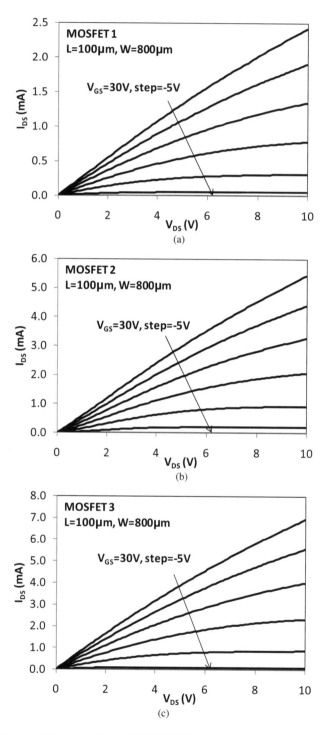

Fig. 3. Output IV characteristics of (a) MOSFET 1, (b) MOSFET 2, and (c) MOSFET 3

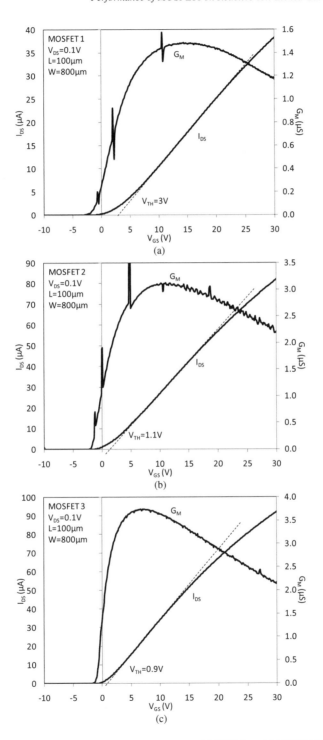

Fig. 4. I_{DS} and G_M vs. V_{GS} of (a) MOSFET 1, (b) MOSFET 2, and (c) MOSFET 3

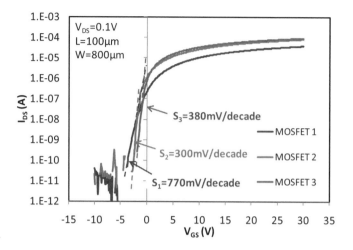

Fig. 5. Log I_D vs. V_G for gate oxide MOSFETs on three different GaN surfaces.

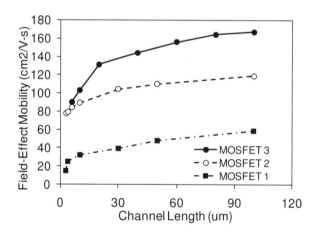

Fig. 6. Field-effect mobility with different channel length for all three MOSFETs.

Moreover, in 100 nm gate oxide devices, a maximum field-effect mobility of 60 cm^2/V-s for MOSFET 1 was extracted on long channel (L_{ch}=100 μm) devices, which are lower than the values of 120 cm^2/V-s in MOSFET 2, and 170 cm^2/V-s in MOSFET 3. These are shown in Fig. 6.

All of the IV characteristics, transconductance, on-resistance, subthreshold slope and field-effect mobility indicate the 1000°C annealing condition applied in MOSFET 2, 3 process is better than the 900°C annealing condition applied in MOSFET 1 process. This trend is consistent with the measurement results of interface densities in MOS capacitors, which has the lowest value with 1000°C annealing temperature [2]. In Table 1, we have summarized the maximum transconductance, on-resistance, subthreshold slope and field-effect mobility of all three MOSFETs.

Table 1. Subthreshold slope and field-effect mobility of MOSFETs on different GaN surfaces.

Sample	Maximum Transconductance (μS)	On-resistance (Ω-mm)	Gate Ox Subthreshold (mV/decade)	Field Ox Subthreshold (V/decade)	Field-effect Mobility (cm^2/V-s)
Etched MOSFET 1	1.49	2944	770	4.8	60
Etched MOSFET 2	3.12	1333	300	1.5	120
As-grown MOSFET 3	3.74	970	380	-	170

4. Summary

We have compared the electrical characteristics of dry/wet etched and un-etched GaN MOSFETs with 900°C and 1000°C gate oxide annealing conditions. Higher on-state current level, higher transconductance, lower on-resistance, lower subthreshold slope and higher field-effect mobility suggest that the 1000°C annealing temperature for the PECVD gate oxide in our MOSFET process is better than 900°C.

Acknowledgement

This work was supported by NSF Center for Power Electronics Systems (EEC-9731677) and Furukawa Electric Co.

References

1. W. Huang, Z. Li, T. P. Chow, Y. Niiyama, T. Nomura and S. Yoshida, "Enhancement-mode GaN Hybrid MOS-HEMTs with $R_{on,sp}$ of 20 mΩ-cm^2," in *Proc. Int. Symp. Power Semicond. Devises ICs*, May 18-22, 2008, pp. 295-298.
2. W. Huang, T. Khan, and T.P. Chow, "Comparison of MOS capacitors on N and P type GaN," *J. Electronic Materials*, vol. 35, pp. 726-732, 2006.
3. W. Huang, T. Khan, and T. P. Chow, "Enhancement-Mode n-Channel GaN MOSFETs on p and n⁻ GaN/Sapphire Substrates," *IEEE Electron Device Lett.*, vol. 27, no. 10, pp. 796-798, Oct. 2006.
4. W. Huang, and T. P. Chow, "Monolithic High-Voltage GaN MOSFET/Schottky Pair with Reverse Blocking Capability," in *Proc. Int. Symp. Power Semicond. Devises ICs*, May 27-31, 2007, pp. 265-268.
5. K. Tang, W. Huang, and T. P. Chow, "GaN MOS Capacitors and FETs on Plasma-Etched GaN Surfaces", accepted by *J. Electronic Materials*.

International Journal of High Speed Electronics and Systems
Vol. 19, No. 1 (2009) 129–135
© World Scientific Publishing Company

HIGH CURRENT DENSITY/HIGH VOLTAGE AlGaN/GaN HFETs ON SAPPHIRE

JUNXIA SHI

School of Electrical Engineering, Cornell University
Ithaca, NY 14853, USA
js544@cornell.edu

M. POPHRISTIC

Velox Semiconductor Corporation
Somerset, NJ 08873, USA

L. F. EASTMAN

School of Electrical Engineering, Cornell University
Ithaca, NY 14853, USA

AlGaN/GaN heterojunction field effect transistors (HFETs) on sapphire substrates for power-switching applications have been fabricated. Design parameters such as source-gate spacing (L_{sg}), gate length (L_g), and gate width (W_g) have been varied to check their effects on the device performances. For a gate-drain spacing (L_{gd}) of 10μm, a specific on-resistance (AR_{on}) of 1.35mΩ-cm^2 and off-state breakdown voltage (BV) of 770V was achieved, which is close to the 4-H SiC theoretical limit.

Keywords: AlGaN/GaN; HFETs

1. Introduction

GaN-based heterojunction field-effect transistors (HFETs) are attractive candidates for low-loss and high-power switching applications, thanks to their high sheet carrier density (n_s) in the 2-dimensional electron gas (2DEG) channel and large critical electric field. Off-state drain-source breakdown voltages (BV) and specific on-resistance (AR_{on}) are among the most important parameters for evaluating power switching devices. Power device figure of merit (FOM), BV^2/AR_{on}, has been widely used to directly compare devices with different BV and AR_{on} ratings from the material property point of view. Compared to SiC power devices, GaN-based HFETs have much lower on-resistance and higher BV. Up to now, on 4-H SiC substrates, a record BV of ~1900V has been achieved; on sapphire substrates, a record BV of ~1600V has been obtained.[1,2] GaN buffer layer resistivity is of great importance because of its role in the minimization of buffer leakage current and the realization of large off-state high-voltage blocking.

In this work, highly resistive C-doped GaN buffer layers grown on 2-in. c-plane sapphire substrates were employed for the fabrication of the high-power switching GaN-based HFETs. The material structure started with a 30nm of AlN nucleation layer on a c-plane sapphire substrate, on top of which 1.8μm C-doped GaN, 1nm of AlN, and 20nm of un-intentionally doped (UID) $Al_{0.28}Ga_{0.72}N$ were grown subsequently by MetalOrganic chemical vapor deposition. Room temperature C-V measurements showed a n_s of about $9.2 \times 10^{12} cm^{-2}$ in the 2-DEG channel and the carrier concentration of the C-doped GaN buffer layer was lower than $1 \times 10^{13} cm^{-3}$.

2. Experimental

The schematic cross-sectional view of the fabricated 2-finger T-shaped AlGaN/GaN HFETs is shown in Figure 1. Cr/Pt was first put down on the chip as the stepper alignment mark. Thereafter, mesa isolation was performed in an ICP-RIE etcher using a chlorine-based gas mixture ($Cl_2/BCl_3/Ar$). Next, Ta/Ti/Al/Mo/Au was evaporated as the source and drain ohmic contacts. Prior to the ohmic metallization, the surfaces were treated with $SiCl_4$ plasma followed by a buffered oxide etch (BOE) to remove any surface oxide layer. Afterwards, rapid thermal annealing (RTA) was performed at 600°C for 5 minutes. On-chip transfer length measurements showed an ohmic contact transfer resistance of 0.8Ω-mm and a sheet resistance of $420\Omega/\square$. Thereafter, a 400nm thick oxide insulation layer was deposited on the mesa-etched region at a substrate temperature of 400°C in a plasma-enhanced chemical vapor deposition (PECVD) system. This insulation layer is to eliminate pad-to-pad current leakage paths through the GaN buffer layer. Then the oxide layer was wet-etched using BOE 6:1. Thereafter, the Ni/Au gates were evaporated and lifted off.

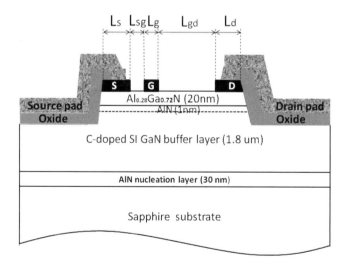

Figure 1 Schematic cross-sectional view of fabricated AlGaN/GaN HFETs (S: source; G: gate; D: drain; L_s/L_d: source/drain contact length; L_{sg}: source-gate spacing; L_g: gate length; L_{gd}: gate-drain spacing).

Previously, the variations of device performances with gate-drain spacing (L_{gd}) and ohmic lengths (L_s, L_d) have been thoroughly studied.[2] In this work, all the fabricated devices have a fixed gate-drain spacing (L_{gd}) of 10μm and fixed ohmic contact lengths ($L_s=L_d=8$μm). Influences of other important design parameters, such as source-gate spacing (L_{sg}), gate length (L_g), and gate width (W_g), on the device characteristics (maximum drain-source current density ($I_{DS,max}$), maximum transconductance ($g_{m,max}$), AR_{on}, and BV) are investigated. L_{sg} and L_g are varied from 1 to 4μm; W_g is varied from 100 to 1000μm.

3. Results and discussions

Figure 2(a) shows the on-state DC I_{DS}-V_{DS} characteristics of the fabricated HFETs with varying L_{sg} and fixed L_g of 1μm and W_g of 200μm. The DC characteristics of the fabricated devices in this work were measured using an HP 4142B modular source/ monitor and microwave probes. V_{GS} was swept from +1 to -4V in steps of -1V. It was shown in Figure 2(a) that under V_{GS} of +1V, $I_{DS,max}$ was ~500mA/mm for the device with L_{sg} of 1μm and it decreased with increasing L_{sg}. For the device with L_{sg} of 4μm, $I_{DS,max}$ is ~420mA/mm, ~15% less than that of the device with L_{sg} of 1μm. In this work, $AR_{DS(on)}$ was defined as the product of $R_{DS(on)}$ measured at $1/2I_{DS,max}$ under $V_{GS}=1$V and A defined by the mesa isolation process. The measured $AR_{DS(on)}$ increased slightly with increasing L_{sg}. For the device with L_{sg} and L_g of 1μm and W_g of 200μm, $AR_{DS(on)}$ was measured to be 1.35mΩ-cm^2. When the device active area (A') was defined only by the region between source and the drain, which does not include the source and drain contacts, the $A'R_{DS(on)}$ was 0.58mΩ-cm^2. The negative resistance on the on-state I_{DS}-V_{DS} characteristics is presumably due to the heating effect as a result of the inferior thermal conductivity of the sapphire substrate. A decrease in $I_{DS,max}$ with increasing L_g was also observed and Figure 2(b) shows the maximum drain-source current density trend with varying total lengths of $L_{sg}+L_g$. Compared to the device with $L_{sg}=L_g=2$μm, there was a ~15% increase in current density for the device with $L_{sg}=L_g=1$μm.

Figure 3 shows the transfer characteristics (I_{DS}-V_{GS} curves, transconductance (g_m)) of the fabricated C-doped SI GaN-on-sapphire HFETs with varying L_{sg} and fixed L_g of 1μm. The devices exhibited excellent pinch-off characteristics. A $g_{m,max}$ of ~200mS/mm was obtained under $V_{DS}=5$V and the V_p was determined to be ~-3.4V. $g_{m,max}$ decreases to ~130mS/mm for the device with L_{sg} of 4μm. A similar trend of transfer characteristics with varying L_g was also observed (not shown). The variation of source-drain current density and transconductance with varying gate width was shown in Figure 4. The decrease in current density with increasing gate width might be due to the potential drop along the source metal feed-line, which literally adds negative bias on the gate. This is currently under investigation.

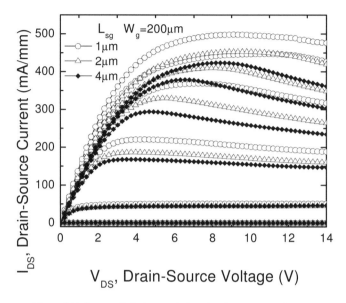

Figure 2(a) On-state DC characteristics of devices with varying L_{sg}.

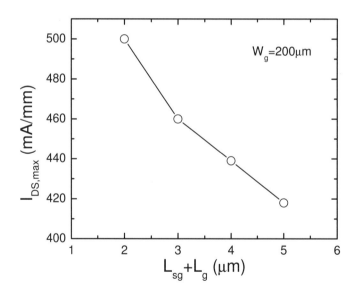

Figure 2(b) Trend of $I_{DS,max}$ with combined length of L_{sg} and L_g.

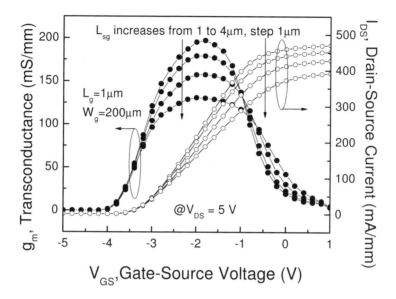

Figure 3 Transfer characteristics of the fabricated devices with varying L_{sg}.

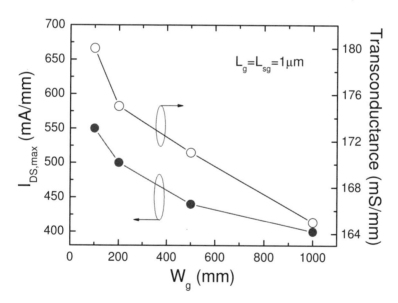

Figure 4 Trend of W_g on $I_{DS,max}$ and g_m.

Breakdown measurements were performed using a Tektronix 370 A curve tracer under V_{GS} of -5.5V. A BV of 770V was obtained from the device with L_{gd} of 10µm. As the average breakdown field was lower than 1MV/cm in this case, which is much lower than the theoretical AlGaN breakdown field strength of ~3 to 5MV/cm, there should be still room for further improvements from both process and material quality points of view. The AR_{on}-BV behavior of the device is shown in Figure 5. For the fabricated GaN HFETs on sapphire substrate, the power device figure of merit ($BV^2/AR_{DS(ON)}$ or $BV^2/A'R_{DS(ON)}$) was about 4.39×10^8 or 1×10^9 V^2/Ω-cm^2, which is close to the 4-H SiC theoretical limit.

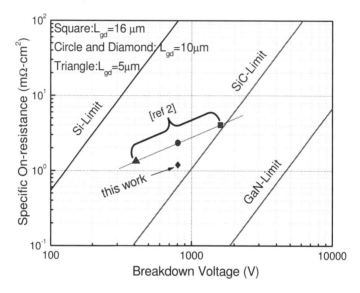

Figure 5 AR_{on}-BV behavior of the fabricated devices ($L_{sg}=L_g=1\mu m$).

4. Conclusion

AlGaN/GaN HFETs with high BV and low AR_{on} were fabricated and characterized on sapphire substrates. The effects of L_{sg}, L_g, and W_g on the device characteristics were investigated. Compared to the device with $L_{sg}=L_g=2\mu m$, the drain-source current density showed an increase of more than ~15%. For the device with L_{sg} and L_g of 1μm, L_{gd} of 10μm, a power device figure of merit ($BV^2/AR_{DS(ON)}$ or $BV^2/A'R_{DS(ON)}$) of 4.39×10^8 or 1×10^9 V^2/Ω-cm^2 was achieved, which is close to the 4-H SiC theoretical limit.

Acknowledgments

This work was performed in part at the Cornell NanoScale Facility, a member of the National Nanotechnoly Infrastructure Network, which is supported by the National Science Foundation (Grant ECS-0335765).

References

1. Y. Dora, A. Chakraborty, L. McCarthy, S. Keller, S. P. DenBaars, and U. K. Mishra, High breakdown voltage achieved on AlGaN/GaN HEMTs with integrated slant field plates, IEEE Electron Device Lett. **27**(9), 713-715 (2006).
2. J. Shi, Y. C. Choi, M. Pophristic, and L. F. Eastman, High breakdown voltage AlGaN/GaN heterojunction field effect transistors on sapphire, *Phys. Stat. Sol.(c)* **5**(6), 2013–2015 (2008).

International Journal of High Speed Electronics and Systems
Vol. 19, No. 1 (2009) 137–144
© World Scientific Publishing Company

InAlN/GaN MOS-HEMT WITH THERMALLY GROWN OXIDE

M. ALOMARI, F. MEDJDOUB, E. KOHN

Institute of Electron Devices and Circuits (EBS), University of Ulm, Albert Einstein Allee 45,
Ulm, 89081, Germany
mohammed.alomari@uni-ulm.de

M-A. DI FORTE-POISSON, S. DELAGE

Alcatel-Thales III-V Laboratory
Marcoussis (Paris), France

J.-F. CARLIN, N. GRANDJEAN

IPEQ-LASPE EPFL
Lausanne (Switzerland), CH 1015 Lausanne, Switzerland

C. GAQUIÈRE

IEMN, U.M.R.-C.N.R.S. 8520, 59652 Villeneuve d'ascq,
Lille, France

We report on the investigation of lattice matched InAlN/GaN MOS-HEMT structures prepared by thermal oxidation at 800 °C in oxygen atmosphere for two minutes. The gate leakage current was reduced by two orders of magnitude. Pulse measurements showed lag effects similar to what is observed for devices without oxidation, indicating a high quality native oxide. The MOS-HEMT showed no degradation in the small signal characteristics and yielded a power density of 5 W/mm at 30 V drain voltage at 10 GHz, power added efficiency of 42% and F_t and F_{max} of 42 and 61 GHz respectively, illustrating the capability of such MOS-HEMT to operate at high frequencies.

Keywords: MOS-HEMT; InAlN/GaN; power; thermal oxidation

1. Introduction

InAlN/GaN heterostructures are a new alternative to AlGaN/GaN HEMT device structures with high 2DEG channel charge density[1] and thermal stability above 1000°C in the lattice matched configuration[2]. High aspect ratio devices with ultra thin barrier layers down to 3 nm have been fabricated without using a gate recess technology[3]. Recently, large signal measurements have resulted in 6.8 W/mm output power at 10 GHz[4]. However, these devices have often suffered from high gate leakage current.

In order to reduce the gate leakage the high thermal stability of the lattice matched InAlN/GaN structure can be employed. Figure 1 shows a 13 nm barrier InAlN/GaN HEMT before and after thermal stress at 1000 °C for 30 minutes in vacuum, where no apparent degradation to the device characteristics occurred. This high thermal stability enables the incorporation of high temperature processes into the device fabrication routine. In this paper we discuss the ability and characteristics of lattice matched InAlN/GaN MOS-HEMT fabricated by thermal oxidation at 800 °C in oxygen atmosphere.

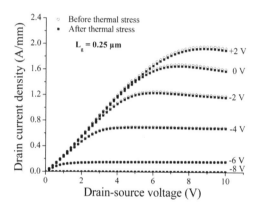

Figure 1: 13 nm InAlN/GaN HEMT before and after thermal stress at 1000 °C for 30 minutes in vacuum. No degradation in the device characteristics was observed after the thermal stress.

2. Growth and processing

The structure studied here was grown by MOCVD system on a 2 inch diameter SiC substrate. The structure consists of 1.7 µm thick GaN buffer, 1 nm thick AlN spacer layer and 10 nm thick InAlN barrier layer with 81% Al content (see fig 2.a). The thin optimized AlN interlayer is used to reduce alloy disorder scattering and thus to improve the transport characteristics[5]. Room temperature Hall effect measurements yielded a 2DEG sheet charge density of 1.6×10^{13} cm^{-2} and a sheet resistivity of 305 $\Omega/\square$.

MOS-HEMT devices were realized as follows: The devices are mesa isolated by Ar plasma etching. Then, Ti/Al/Ni/Au ohmic contact stacks are alloyed at 890°C in nitrogen atmosphere to provide contact resistances of 1.2 Ω.mm, as measured by linear TLM (see fig. 2.a). Blanket thermal oxidation of the whole surface was conducted at 800 °C in oxygen atmosphere for two minutes and 5 minutes (see fig. 2.b). The submicron gates ($L_g = 0.25$ µm) are defined with electron beam lithography. Ni/Au is deposited to produce the gate diode contacts (see fig 2.c). The devices were then passivated by a 100 nm thick Si$_3$N$_4$ layer (see fig 2.d).

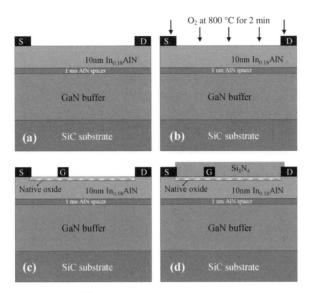

Figure 2: The MOS-HEMT fabrication process. (a) Mesa and ohmic contacts. (b) Blanket thermal oxidation in oxygen at 800 °C. (c) gate lithography and deposition. (d) Si_3N_4 passivation.

3. Experimental results

Figure 3 (left) shows the gate diode characteristics without oxidation and for oxidation times of 2 and 5 minutes. The leakage current was reduced by two orders of magnitude for devices oxidized for 2 minutes and three orders of magnitude for devices oxidized for 5 minutes. The leakage current is in essence reduced exponentially with oxidation time. This indicates that initially leakage is by tunnelling and that this tunnelling current is reduced by an intermediate oxide layer. A square root time dependence indicates also that the oxidation mechanism is diffusion limited (similar to the case of Si) with an initial oxidation rate of approx. 1 nm/min (although final verification can only be obtained by TEM cross sections). Comparing the un-oxidized case with that of 5 min oxidation two effects can be noticed, (despite the fact that the blanket oxidation of the free surface and contacts will introduce parasitic effects): a shift in threshold voltage and a reduction in open channel current (see figure 3). Both are a clear indication of a thinner barrier with higher surface depletion due to a residual surface potential[6]. The MOSHEMT structure investigated here contains therefore an oxide barrier in the order of the tunnelling thickness and the MOS diode is therefore still barrier controlled and may also be labelled a dielectric assisted Schottky contact. But the oxide barrier is clearly visible through the shift in the onset of forward gate leakage. However, in advanced FET structures high aspect ratios and nm-gatelengths are required. In the case of InAlN barriers scaling has been successful down to 3 nm, which is also at the tunnelling limit. Thus also oxide barriers of comparable thickness are required not to degrade the structural aspect ratio of the FET.

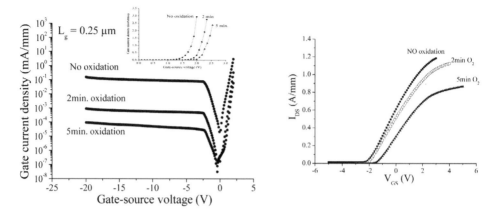

Figure 3: Gate diode characteristics after various thermal oxidation times at 800 °C (left) and transfer characteristics (right).

Pulse experiments were performed on MOS-HEMTs oxidized for 2 minutes to asses the stability of the device with 500 ns pulse duration and 10 μs intervals (see fig. 4). All quiescent bias points (V_{DS0}, V_{GS0}) are chosen in order to simultaneously eliminate the thermal effect (cold polarization) and to reveal the gate and drain lag effects: ($V_{DS0} = 0$ V, $V_{GS0} = 0$), ($V_{DS0} = 0$ V, $V_{GS0} = -3$) and ($V_{DS0} = 20$ V and 25 V , $V_{GS0} = -3$).

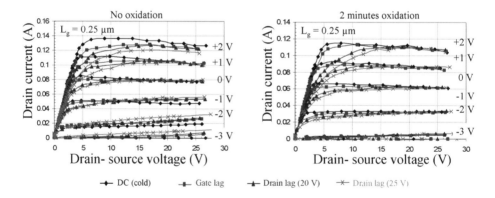

Figure 4: Pulsed I_D-V_{DS} characteristics (V_{GS} swept from -3 to 2 V by step of 2 V) at the quiescent bias points ($V_{DS0} = 0$ V, $V_{GS0} = 0$), ($V_{DS0} = 0$ V, $V_{GS0} = -3$) and ($V_{DS0} = 20$ V and 25 V, $V_{GS0} = -3$) for HEMT (left) and MOS-HEMT (right).

These measurements showed very little gate lag effects, indicating a low trap density in the MOS gate diode. Drain lag was also typical and comparable to that of devices without oxidation. An indication maybe that the saturated output power density of the MOSHEMT was 5.0 W/mm (see figure 5), as compared to 6.8 W/mm in the un-oxidized case as discussed below. The difference may indeed still indicate a higher trap density in the lateral gate to drain high field drift area, which may on the other hand be related to the Si_3N_4 passivation on top of the oxidized surface.

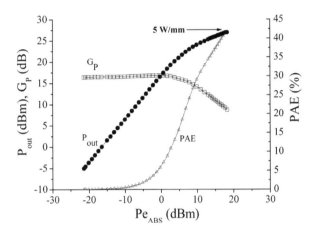

Figure 5: Load-pull measurements on 0.25 μm x 50 μm MOS-HEMT. The device yielded a power density of 5 W/mm and 42% power added efficiency.

The s-parameter measurements (see fig. 6) of 0.25 μm x 50 μm MOS-HEMT (2min oxidation, no T-Gate) showed a cut-off frequency (F_t) of 42 GHz and a maximum oscillation frequency (F_{max}) of 61 GHz compared to F_t of 37 GHz and F_{max} of 60 GHz for HEMT device of the same gate length without oxidation, indicating no degradation of the small signal channel transport properties. Load pull measurements up to 30 V drain bias have resulted in up to 5 W/mm saturated power density at 10 GHz along with a power added efficiency of 42% (see fig. 5). To our knowledge, this is the first time that such power measurements could be performed on thermally oxidized nitride based MOSHEMT structures.

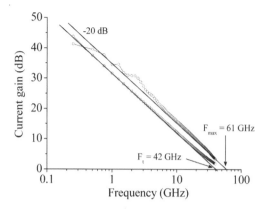

Figure 6: Frequency performances (right) of a 0.25×50 μm² 10 nm barrier In$_{0.18}$Al$_{0.82}$N/GaN MOSHEMT on SiC substrate.

4. Conclusion and outlook

We have investigated the potential of realizing InAlN/GaN MOS-HEMT structures using a thermal oxidation process in order to reduce the gate leakage of the device. Although thermal oxides have not been suitable for MOSFET application in the case of III-V semiconductors and also AlGaN/GaN heterostructures, these initial results on InAlN-MOSHEMTs with thermally grown oxide gate dielectric show already promising small signal and large signal characteristics. The gate leakage of the device was indeed reduced by two orders of magnitude for oxidation time of 2 minutes and three orders of magnitude for an oxidation time of 5 minutes. s-parameter and load-pull measurements showed no degradation of the small signal response of the device. Thus, the oxide is of high electronic quality. Load pull measurements yielded a maximum saturated power density of 5 W/mm. However, the drain lag characteristics indicate still trapping in the high field drain drift region of the device. It is thought that this is related to the passivation with Si$_3$N$_4$ and the oxide/Si$_3$N$_4$ interface.

In order to avoid unnecessary oxidation of the metal contacts and the free surface areas between source and gate, and gate and drain respectively, the thermal oxidation process can be confined to the gate area only (see fig. 7). As a first proof-of-concept the oxidation process has been applied to a self-aligned gate fabrication routine of a 10 nm InAlN/GaN layer grown on sapphire, resulting in a self-aligned MOS-HEMT configuration. The DC output characteristics of such a device are shown in figure 8. Work is underway to optimize such structures.

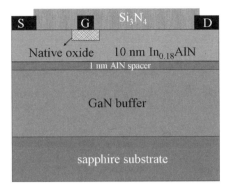

Figure 7: Cross section of InAlN/GaN MOS-HEMT prepared by thermal oxidation at 800 °C where the oxide is localized only under the gate area.

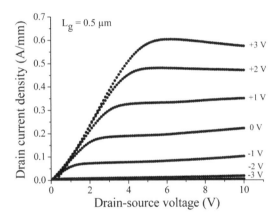

Figure 8: DC output characteristics of 0.5 µm x 50 µm InAlN7GaN MOS-HEMT with the oxidation process localized under the gate area only.

5. Acknowledgements

This work was carried within the frame of the European Union Project (UltraGaN), contract #6903.

References

1. J. Kuzmik *et al*, "Power electronics on InAlN/(In)GaN: Prospect for a record performance," *IEEE Electron Device Lett.*, Vol. 22, p. 510, 2001
2. F. Medjdoub *et al* "Can InAlN/GaN be an alternative to high power/high temperature AlGaN/GaN devices?," *IEDM Tech. Dig.*, p. 927, 2006.
3. F. Medjdoub *et al* "Thermal stability of extremely thin barrier InAlN/GaN HEMTs" *ISDRS.*, College Park (MD), 2007.

4. N. Sarazin *et al* "X-band power characterisation of AlInN/AlN/GaN HEMT grown on SiC substrate," *Electronics Lett.*, 43 (23), 2007.

5. J. –F. Carlin *et al* "Improved AlInN/GaN high electron mobility transistor structures" *IWN 2006*, Kyoto, Japan, 2006

6. F. Medjdoub *et al* "Barrier layer downscaling of InAlN/GaN HEMTs", *Device Research Conference*, Indiana (USA), DRC Tech. Dig., p. 109, *2007*.

International Journal of High Speed Electronics and Systems
Vol. 19, No. 1 (2009) 145–152
© World Scientific Publishing Company

GaN TRANSISTORS FOR POWER SWITCHING AND MILLIMETER-WAVE APPLICATIONS

TETSUZO UEDA, YASUHIRO UEMOTO, TSUYOSHI TANAKA, and DAISUKE UEDA[*]

Semiconductor Device Research Center, Semiconductor Company, Panasonic Corporation
1 Kotari-yakemachi, Nagaokakyo-shi, Kyoto 617-8520, JAPAN
ueda.tetsuzo@jp.panasonic.com

We review our state-of-the-art GaN-based device technologies for power switching at low frequencies and for future millimeter-wave communication systems. These two applications are emerging in addition to the power amplifiers at microwave frequencies which have been already commercialized for cellular base stations. Technical issues of the power switching GaN device include lowering the fabrication cost, normally-off operation and further increase of the breakdown voltages extracting full potential of GaN-based materials. We establish flat and crack-free epitaxial growth of GaN on Si which can reduce the chip cost. Our novel device structure called Gate Injection Transistor (GIT) achieves normally-off operation with high enough drain current utilizing conductivity modulation. Here we also present the world highest breakdown voltage of 10400V in AlGaN/GaN HFETs. In this paper, we also present high frequency GaN-based devices for millimeter-wave applications. Short-gate MIS-HFETs using in-situ SiN as gate insulators achieve high f_{max} up to 203GHz. Successful integration of low-loss microstrip lines with via-holes onto sapphire enables compact 3-stage K-band amplifier MMIC of which the small-signal gain is as high as 22dB at 26GHz. The presented devices are promising for the two future emerging applications demonstrating high enough potential of GaN-based transistors.

Keywords: AlGaN/GaN; heterojunction FET; Gate Injection Transistor; on-state resistance; breakdown voltage; MIS-HFET; microstrip line

1. Introduction

AlGaN/GaN heterojunction field effect transistors (HFETs) have been widely investigated primary for power amplifiers at microwave frequencies to be used for wireless communications taking advantage of the superior material properties. So far, these power amplifiers mainly for cellular-phone base stations have been commercialized which should make the system highly efficient and very compact. Besides such microwave applications, high power switching at low frequencies and further extension of operating frequencies to millimeter-wave range are promising as future emerging applications of GaN-based devices. In this paper, we review our state-of-the-art device technologies for such two future applications of GaN transistors.

[*] Present affiliation: *Advanced Technology Research Laboratories, Panasonic Corporation, 3-4 Hikaridai, Seika-cho, Soraku-gun, 619-0237 Kyoto, JAPAN.*

2. GaN Transistors for Power Switching

GaN-based power switching transistors have been expected to replace the currently used Si-based MOSFETs and IGBTs enabling highly efficient switching owing to the low on-state resistance and the high breakdown voltage. Lowering the fabrication costs of these GaN devices is critical for the widely-spread use of them, and so far the cost of the substrate for epitaxy such as SiC has been dominant in the total costs. We establish the epitaxial growth of GaN on cost-effective Si using AlN/GaN superlattice interlayer which relaxes the epitaxial strain caused by the lattice and thermal mismatches. Crack-free and mirror surfaces of AlGaN/GaN are obtained over 6-inch diameter Si substrate as shown in Fig.1, where the highest electron mobility of 1653cm^2/Vsec is confirmed. Using the epitaxial growth technologies, we have demonstrated high enough performance of AlGaN/GaN power switching transistors on Si substrates[1].

The most crucial task left for the GaN transistors is to achieve normally-off operation which is strongly desired for the safety operation, however, it has been very difficult because of the built-in polarization electric field in the GaN-based material systems. Our solution for the normally-off operation is a new device called GIT (Gate Injection Transistor) as shown in Fig.2(a)[2]. The GIT features the p-AlGaN gate formed over the undoped AlGaN/GaN hetero-structure. The p-AlGaN lifts up the potential at the channel, which results in normally-off operation as shown in the band diagram of Fig.2(b). Fig.3 illustrates the

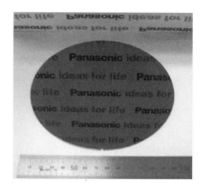

Fig. 1 Photograph of a crack-free AlGaN/GaN epitaxial wafer on a 6-inch diameter Si substrate.

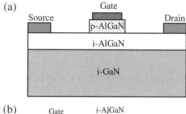

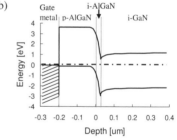

Fig. 2 A schematic illustration of
(a) GIT structure and
(b) calculated band diagram of GIT at the gate bias of 0V. The epitaxial structure is designed to serve normally-off operations.

basic operation of the GIT utilizing conductivity modulation. At the gate voltages up to the forward built-in voltage V_f of the pn junction at the gate, the GIT is operated as a field effect transistor. Further increase of the gate voltage exceeding the V_f results in the hole injection to the channel from the p-AlGaN. The injected holes generate the equal numbers of electrons to keep charge neutrality at the channel. The generated electrons are moved by the drain bias with high mobility, while the injected holes stay around the gate because the hole mobility is two order of magnitude lower than that of the electron. This results in significant increase of the drain current keeping the low gate current. Fig.4

shows the I_{ds}-V_{gs} and g_m-V_{gs} characteristics of the fabricated GIT comparing with those of the conventional Schottky-gate FET. The GIT exhibits peculiar transconductance characteristics with two peaks. Followed by the FET operation up to the V_{gs} of 3V, the I_{ds} is super-linearly increased corresponding to the second g_m peak, which is a direct evidence of the conductivity modulation caused by the hole injection.

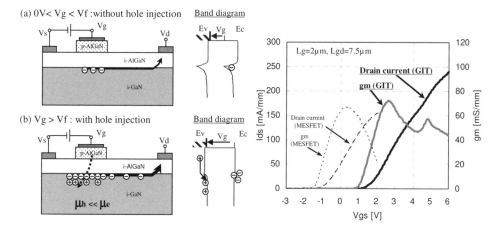

Fig. 3 A schematic illustration of GIT operation varying gate voltages :
(a) without hole injection at Vg < Vf,
(b) with hole injection at Vg > Vf.

Fig. 4 Ids-Vgs abd gm-Vgs characteristics of fabricated (a) GIT and (b) MESFET.

Another technical task for GaN power switching transistor is to further increase the breakdown voltage extracting the full advantage of the superior material properties of GaN. So far, the reported highest off-state breakdown voltage of AlGaN/GaN HFETs maintaining low on-state resistance has remained 1900V at highest[3]. Possible reason of such low value is surface flashover in the air or breakdown of passivation with low dielectric strength, which occurs between electrodes in the lateral device configuration[4]. Overcoming these issues, we have demonstrated ultra high breakdown voltages of AlGaN/GaN HFETs with a novel structure avoiding the undesired breakdown of which the schematic cross section is shown in Fig.5[5]. The most notable feature of the HFET is the via-hole through sapphire at the drain electrode of which the peripheral length can be designed as small as possible. Placing the small drain electrode at the center of the concentric unit cell of the transistor effectively suppresses the premature breakdown between the surface electrodes. A closely-packed array of the unit cell enables a very efficient chip layout as well. Since conventional wet or dry etching cannot be used for chemically stable sapphire, high power pulsed laser is used to form the via-holes. Thick gold plating, wafer thinning and backside metallization follow the laser drilling of sapphire. The thick AlN passivation also helps to increase the breakdown voltages owing to the high breakdown strength. The AlN is deposited by DC-sputtering technique which has 200 times higher thermal conductivity than that of conventional SiN and thus effectively reduces the thermal resistance of the HFET[6]. The optical microscopic image and cross sectional SEM image of the fabricated HFET are shown in Fig. 6 and Fig. 7, respectively.

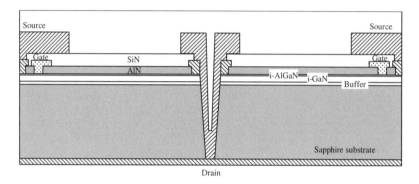

Fig. 5 A schematic cross-section of the fabricated ultra high voltage HFETs with via-holes through sapphire and AlN passivation.

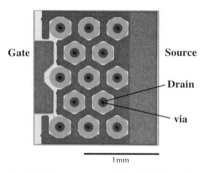

1mm

Fig. 6 Microphotograph of the fabricated ultra high voltage HFET.

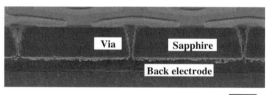

100μm

Fig. 7 Cross sectional SEM image of fabricated via-holes through sapphire.

Fig. 8 shows the resultant off-state breakdown voltages (BV$_{ds}$) as a function of the L$_{gd}$ for various passivations. The BV$_{ds}$ of the HFET with AlN passivation, proportionally increases as L$_{gd}$ increases, while the BV$_{ds}$ saturates at 3000V in the HFETs with SiN passivation. The obtained 10400V with L$_{gd}$ of 125μm is the highest value ever reported for GaN-based transistors, which also shows a

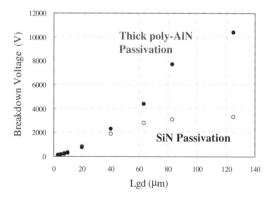

Fig. 8 The off-state breakdown voltages (BV$_{ds}$) of fabricated HFETs for various L$_{gd}$. BV$_{ds}$ of 10400V at Lgd=125μm is obtained with AlN passivation.

strong sign of further increase with the extension of L_{gd} over 125μm. The specific on-state resistance $R_{on} \cdot A$ is 186mΩ·cm^2 with the I_{max} of 150mA/mm for the ultra high voltage AlGaN/GaN HFET with L_{gd} of 125μm. The above two results of the normally-off GIT and the ultra high voltage HFET are plotted in Fig. 9 comparing those of the state-of-the-art Si-based power devices. The presented devices demonstrate that GaN is advantageous in wide range of the operating voltages for various applications.

3. GaN HFETs/MMICs for Future Millimeter-wave Communications

GaN devices are also promising for millimeter-wave frequency range since the material system should be an only viable

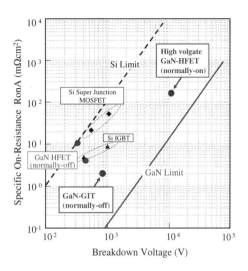

Fig.9 $R_{on} \cdot A$ and breakdown voltage of the fabricated Gate Injection Transistor and ultra high voltage AlGaN/GaN HFET comparing with those of state-of-the-art GaN-based and Si-based devices.

choice for high power amplifier at such high frequencies to increase the distance of the wireless communication. In addition, the GaN device is advantageous in the receiver devices in the millimeter-wave communication systems, because the robust devices can be operated under harsh environments typically at around the outside antenna. The inherent low noise feature is attractive as well[7]. Fig.10 shows the schematic cross section of our state-of-the art AlGaN/GaN MIS-HFET with short gate length down to 100nm[8]. The in-situ SiN is formed subsequently after the MOCVD in the same reactor without any exposure in the air. The SiN exhibits crystalline structure with abrupt interfaces as shown in Fig. 11, by which sheet carrier concentration at the 2DEG isdramatically

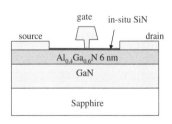

Fig. 10 Schematic cross section of the fabricated AlGaN/GaN MIS-HFET.

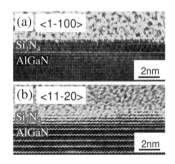

Fig. 11 Cross sectional TEM images at the in-situ SiN/AlGaN interface. The zone axis of (a) <1-100> and (b) <11-20> directions are shown.

increased[9]. Fig.12 shows the measured small signal RF characteristics of the MIS-HFET, where the current gain cut-off frequency (f_T) of 71 GHz and the maximum oscillation frequency (f_{max}) of 203 GHz are obtained keeping the high enough off-state breakdown voltage at around 200V. These results indicate that GaN transistors are also promising for millimeter-wave applications.

In order to implement the GaN devices for practical use especially as the receiving device, transmission lines need to be integrated. Sapphire is an attractive choice of a substrate for the GaN-based MMIC since its insulating nature would greatly help to reduce the loss of the transmission line. We demonstrate a compact 3-stage K-band GaN-based MMIC with integrated microstrip lines with high gain of which the chip photograph is shown in Fig 13[10]. The via-holes for the microstrip lines are successfully formed onto sapphire by the laser drilling technique. Note that the fully integrated matching networks were designed to give the gain matching for the input and output impedances of the three HFETs The measured small signal gain S_{21} of the 3-stage amplifier are shown in Fig. 14, where a small-signal gain as high as 22dB at 26GHz with a 3dB bandwidth of 25-29GHz is confirmed. The presented configuration of the MMIC would be applicable to future millimeter-wave communication systems to be used under harsh environments taking advantage of the robust nature of the GaN-based material systems.

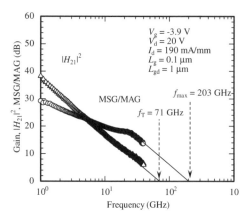

Fig. 12 Small signal RF characteristics of the AlGaN/Gan MIS-HFET.

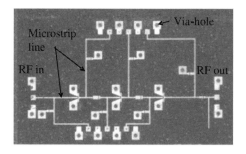

Fig. 13 A chip photograph of the fabricated K-band MMIC.

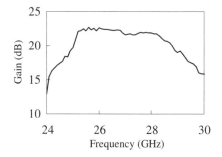

Fig. 14 Measured small signal gain of the 3-stage amplifier.

4. Conclusions

In this paper, we review our state-of-the-art GaN-based device technologies for power switching and future millimeter-wave communications. As for the power switching GaN devices, we present a novel device structure called GIT, which enables normally-off operation with high drain current utilizing conductivity modulation. Here we also present the world highest breakdown voltage of 10400V in AlGaN/GaN HFETs. The device structure with via-holes through sapphire and thick AlN passivation helps to eliminate undesired breakdown between electrodes via passivation dielectric. In the last part of this paper, we present GaN-based MIS-HFETs with in-situ SiN as gate insulators, which exhibit as high f_{max} as 203GHz. The successful integration of low-loss microstrip lines with via-holes onto sapphire enables compact 3-stage K-band amplifier MMIC. The MMIC exhibits a small-signal gain as high as 22dB at 26GHz with a 3dB bandwidth of 25-29GHz. The presented devices are promising for the two future emerging applications demonstrating high enough potential of GaN-based transistors.

Acknowledgements

The authors would like to express thanks to Dr. Kaoru Inoue, Mr. Hiroyuki Sakai, Dr. Manabu Yanagihara, Dr. Hidetoshi Ishida, Mr. Masaaki Nishijima, Dr. Toshiyuki Takizawa, Dr. Shuichi Nagai, Mr. Masahiro Hikita, Dr. Tomohiro Murata, Mr. Satoshi Nakazawa, Dr. Hiroaki Ueno, Mr. Hisayoshi Matsuo, Mr. Masayuki Kuroda, and Mr. Daisuke Shibata in Semiconductor Device Research Center, Matsushita, for their dedicated research works on GaN electronic devices. The authors also would like to thank Prof. Takashi Egawa, Nagoya Institute of Technology, for his valuable technical advices on the epitaxial growth of GaN on Si through the collaborative research.
This work is partially supported by "The research and development project for expansion of radio spectrum resources" of the Ministry of Internal Affairs and Communications, Japan.

References

1. M. Hikita, M. Yanagihara, K. Nakazawa, H. Ueno, Y. Hirose, T. Ueda, Y. Uemoto, T. Tanaka, D. Ueda, and T. Egawa, "350V/150A AlGaN/GaN power HFET on Silicon substrate with source-via grounding (SVG) structure," *IEEE Trans. Electron Device*, vol.52, no.9, pp1963-1968, 2005.

2. Y. Uemoto, M. Hikita, H. Ueno, H. Matsuo, H. Ishida, M. Yanagihara, T. Ueda, T. Tanaka, D. Ueda, "Gate Injection Transistor (GIT)—A Normally-Off AlGaN/GaN Power Transistor Using Conductivity Modulation,"*IEEE Trans. Electron Device*, vol.54, no.12, pp3393-3399, 2007.

3. Y. Dora, A. Chakraborty, L. McCarthy, S. Keller, S. P. DenBaars, and U. K. Mishra, "High-Breakdown Voltage Achieved on AlGaN/GaN HEMTs With Integrated Slant Field-Plate", *IEEE Electron Device Letters*, vol.27, no.9, pp713-715, 2006.

4. N.Tipimeni, V.Adivarahan, G.Simin, and A.Khan, "Silicon Dioxide-Encapsulated High-Voltage AlGaN/GaN HFETs for Power-Switching Applications", IEEE Electron Device Letters, vol.28, no.9, pp784-786, 2007.

5. Y. Uemoto, D. Shibata, M. Yanagihara, H. Ishida, H. Matsuo, T. Ueda, T. Tanaka, and D. Ueda, "8300V Blocking Voltage AlGaN/GaN Power HFET with Thick Poly-AlN Passivation." *IEDM Technical Digests*, pp. 861-864, December 2007.

6. N. Tsurumi. H. Ueno, T. Murata, H. Ishida, Y. Uemoto, T. Ueda, K. Inoue, and T. Tanaka, "AlN Passivation over AlGaN/GaN HFETs for High Power Operation.", *7th Topical workshop on Heterostructure Microelectronics*, Kisarazu, Chiba, Japan , Aug. 2007.

7. Y. Hirose, Y. Ikeda, M. Ishii, T. Murata, K. Inoue, T. Tanaka, H. Ishikawa, T. Egawa, and T. Jimbo, "Low Noise and Low Distortion Performances of an AlGaN-GaN HFETs", *IEICE Trans. Electron.*, vol.E86-C, pp.2058-2064, 2003.

8. M. Kuroda, T. Murata, S. Nakazawa, T. Takizawa, M. Nishijima, M. Yanagihara, T. Ueda, and T.Tanaka, "High fmax with High Breakdown Voltage in AlGaN/GaN MIS-HFETs Using In-situ SiN as Gate Insulators." *2008 IEEE Compound Semiconductor Integrated Circuit Symposium*, Monterey, CA, U.S.A., October 2008.

9. T. Takizawa, S. Nakazawa, and T Ueda, "Crystalline SiNx Ultrathin Films Grown on AlGaN/GaN Using In-Situ Metalorganic Chemical Vapor Deposition," *J. Electron. Mater.*, vol. 37, pp.628-634, 2008.

10. T. Murata, M. Kuroda, S.Nagai., M. Nishijima, H. Ishida, M. Yanagihara, T. Ueda, H. Sakai, T.Tanaka, and M. Li, "A K-band AlGaN/GaN-based MMIC Amplifier with Microstrip Lines on Sapphire", 2008 *IEEE International Microwave Symposium*, Atlanta, GA, U.S.A., June 2008.

International Journal of High Speed Electronics and Systems
Vol. 19, No. 1 (2009) 153–159
© World Scientific Publishing Company

4-NM AlN BARRIER ALL BINARY HFET WITH SiNx GATE DIELECTRIC

TOM ZIMMERMANN, YU CAO, DEBDEEP JENA, HUILI GRACE XING*

Electrical Engineering Department, University of Notre Dame, Notre Dame, IN 46556, USA†
hxing@nd.edu

PAUL SAUNIER

Triquint Semiconductors, Dallas, TX

Undoped AlN/GaN heterostructures, grown on sapphire by molecular beam epitaxy, exhibit very low sheet resistances, ~ 150 Ohm/sq, resulting from the 2-dimensional electron gas situated underneath a 4 nm thin AlN barrier. This extraordinarily low sheet resistance is a result of high carrier mobility and concentration (~ 1200 cm^2/Vs and ~ 3.5×10^{13} cm^{-2} at room temperature), which is ~ 3 x smaller than that of the conventional AlGaN/GaN heterojunction field effect transistor (HFET) structures. Using a 5 nm SiNx deposited by plasma enhanced chemical vapor deposition as gate-dielectric, HFETs were fabricated using these all binary AlN/GaN heterostructures and the gate tunneling current was found to be efficiently suppressed. Output current densities of 1.7 A/mm and 2.1 A/mm, intrinsic transconductance of 455 mS/mm and 785 mS/mm, were achieved for 2 µm and 250 nm gate-length devices, respectively. Current gain cut-off frequency f_T of 3.5 GHz and 60 GHz were measured on 2 µm and 250 nm gate-length devices, limited by the high ohmic contact resistance as well as the relatively long gate length in comparison to the electron mean free path under high electric fields.

1. Introduction

Aggressive downscaling of gate-lengths into the deep sub-micrometer range for high-frequency power-devices demands very thin barrier heterojunction field effect transistors (HFET) structures with a high 2-dimensional electron gas (2DEG) density.[1] To this end, we have explored all binary AlN/GaN heterostructures with a 2.3 nm - 4.0 nm thin AlN barrier grown by plasma-assisted molecular beam epitaxy (PAMBE).[2,3,4] Due to the strong polarization effects in the III-V nitride material system,[5] a 2DEG forms near the AlN/GaN interface with a sheet carrier density as high as 3.5×10^{13} cm^{-2} and mobility as high as 1200 cm^2/Vs.[6] The resultant sheet-resistance is as low as 150 Ω/sq, among the lowest ever reported and ~ 3X smaller than that of the conventional AlGaN/GaN HFETs. Recently we have demonstrated the highest reported output current density of 2.3 A/mm and the highest intrinsic transconductance of 1 S/mm simultaneously,[7] based on similar AlN/GaN all binary heterostructures with, however, a rather poor Al_2O_3 gate dielectric deposited by electron beam evaporation. In this work we have employed a 5.0 nm thin SiNx as the gate dielectric (Fig. 1) deposited by plasma-enhanced chemical vapor deposition (PECVD). This SiN_x film was found to be uniform and dense thus gate leakage in these HFETs is successfully suppressed.

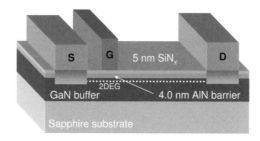

Fig. 1. Schematic cross-section of an AlN/GaN HFET with a 5.0 nm thin SiN_x gate-dielectric on top of a 4.0 nm thin AlN barrier.

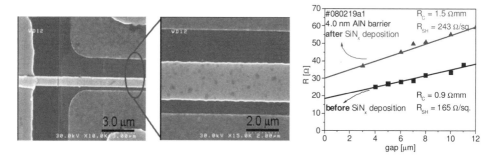

Fig. 2. (Left) SEM images showing smooth annealed ohmic contacts. Gate is the middle line. (Right) TLM measurements showing that the annealed ohmic contact resistance increased to 1.5 ohm-mm after SiN_x deposition.

2. Experiments

The AlN/GaN heterostructures, consisting of a 4.0 nm undoped AlN barrier and a ~ 200 nm thick unintentionally doped (UID) GaN buffer layer, were grown on semi-insulating GaN-on-sapphire substrate in a Veeco 930 PAMBE system. The Hall effect measurements showed a 2DEG density of 3.5×10^{13} cm^{-2} with an electron mobility of 1185 cm^2/Vs at room temperature. As a result, a very low sheet resistance of about 150 Ω/square was achieved, among the lowest ever reported for a single heterostructure. The details of the material growth and transport study have been reported elsewhere.[8] The device mesa (~ 110 nm) was first formed by a Cl_2/BCl_3 based reactive ion etch (RIE). The Ti/Al-based ohmic metal stack was then deposited by electron beam evaporation, followed by a rapid thermal annealing at 570°C resulting in contact resistances of 0.9 Ωmm. The more detailed study on ohmic contact formation to the ultrathin AlN/GaN structures can be found elsewhere.[9] Subsequently a 5 nm thin PECVD-SiN_x layer was deposited and smooth surface morphology and complete coverage of SiN_x was confirmed by atomic force microscopy with a rms-roughness of 0.446 nm. After a second deep mesa etch (~ 250 nm) and a field-oxide deposition, Ni/Au gate metals were deposited on top of the SiN$_x$/AlN/GaN structure. Finally Ti/Au pads were deposited. Secondary electron microscopy (SEM) images (Fig. 2) show the ultrasmooth ohmic contacts, which in turn allows us to place the gate as close as possible to the source contacts. The transmission line method (TLM) measurements after the device completion showed an increase of contact resistances to be ~ 1.5 Ωmm (Fig. 2), which is currently ascribed to the plasma damage to the 2DEG during SiN$_x$ deposition.

3. Results and Discussion

A good rectifying behavior of the SiN$_x$-insulated gate-diode for all measured gate-lengths was observed and the current levels were in the range of μA/mm at an applied voltage in excess of 15 V. Thus, the gate-tunnel-current due to the thin AlN barrier was effectively suppressed by introducing the 5 nm SiN$_x$ layer. Fig. 3 shows the transfer curve of the drain output current and gate leakage current measured simultaneously on a 2 μm long device. A subthreshold slope of ~ 0.65 V/dec was also extracted (Fig. 3 inset), which is clearly limited by the relatively high buffer leakage current. The carrier concentration profile was extracted by C-V measurements, shown in the right of Fig. 3. A 2DEG is clearly observed at the AlN/GaN interface and a rather large number of charges can also be seen near the GaN regrowth interface, confirming the source of the high buffer leakage current observed in these devices.

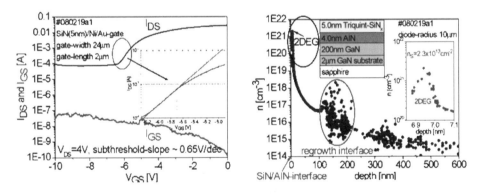

Fig. 3. (Left) Logarithmic plot of drain-output current and gate-leakage current vs. the gate-source voltage of a 2 μm gate-length FET device indicates the pinch-off current level is not limited by gate-leakage current. A high subthreshold slope of 0.65 V/dec and the low gate-leakage current point toward some parasitic contribution by the buffer (inset). (Right) Carrier concentration versus depth determined by CV measurements. A rather large number of charge was observed near the GaN regrowth interface.

Owing to the finite parallel conduction at the regrowth interface, a deep mesa etch was introduced to fully isolate devices from each other. Fig. 4 shows the family I-Vs of the 2 μm long gate HFET and the gate leakage current simultaneously measured. A maximum DC output current density of 1.7 A/mm was achieved. The contribution of the gate leakage current to the output current density is negligible since it is about three orders of magnitude smaller than the output current. Furthermore, it can be seen from Fig. 3 that the contribution from the buffer leakage to the output current is about 100X smaller than the contribution from the 2DEG. Therefore, we can conclude that the output current of these devices indeed stem from the 2DEG at the AlN/GaN heterojunctions.

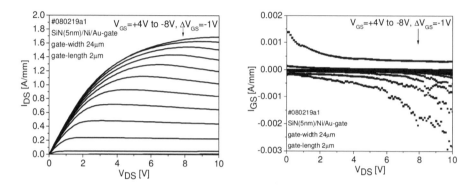

Fig. 4. (Left) DC output characteristics of a 2 μm gate-length HFET with a maximum output current density of 1.7 A/mm. (Right) The measured gate current leakage is three orders of magnitude smaller than the output current and does not contribute substantially to the output current level.

The same 2 μm gate-length device shows an extrinsic transconductance $g_{m,ext}$ of 270 mS/mm at V_{DS}=4V and the transfer-curve indicates a threshold voltage of -5.5 V (Fig. 5). Since the ohmic contact resistance R_c is high in these devices, it is interesting to extract the intrinsic transconductance thus understanding the potential of these devices when the ohmic contact resistance can be sufficiently low. Using R_c=1.5 ohm-mm, $g_{m,int}$ of 455 mS/mm is extracted, which is among the highest reported in AlGaN/GaN based HFETs with 2 μm gate length.

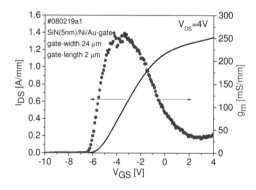

Fig. 5. Transfer-characteristic and transconductance of a 2 μm gate-length HFET showing a maximum $g_{m,ext}$ of 270 mS/mm (corresponding to $g_{m,int}$ of 455 mS/mm).

HFETs with gate-length of 250 nm were also characterized, showing a DC output current of ~ 2.1 A/mm and an extrinsic transconductance of ~ 360 mS/mm (corresponding to $g_{m,int}$ = 785 mS/mm), comparable to what we have reported previously [Ref.7]. Shown in Fig.6 are the small signal characteristics of 250 nm gate devices, measured using an Agilent 8722D network analyzer. A maximal current cutoff frequency f_T of 46 GHz and a power cutoff frequency $f_{max(MAG)}$ of 50 GHz were determined. After de-embedding to remove the influence of the parasitic pad capacitances, a marginal increase in the cutoff frequency was observed, e.g. f_T increased to 60 GHz.

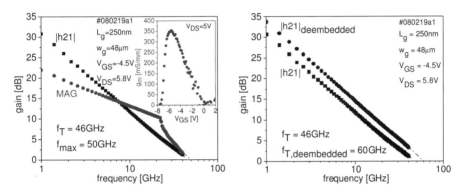

Fig. 6. (Left) Current cutoff frequency f_T of 46 GHz and power cutoff frequency f_{max} of 50 GHz were measured for a 250 nm gate-length FFET with g_m of 360 mS/mm. (Right) After de-embedding f_T of 60 GHz was extracted.

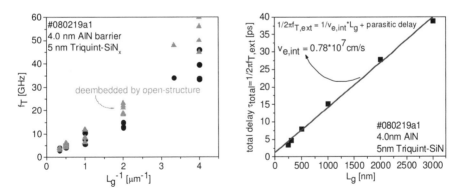

Fig. 7. (Left) As-measured and de-embedded values of f_T versus HFET gate-length (250 nm – 3 µm). (Right) An average electron velocity of ~ 0.78 x 10^7 cm/s extracted from the deembedded f_T values assuming the parasitic delay is the same for all devices.

The as-measured and de-embedded current cutoff frequency values are summarized in Fig. 7 with respect to the gate length. Assuming the contribution of parasitic elements to the total time delay is the same for all HFETs with various gate lengths, it is found that the average electron velocity is ~ 0.78 x 10^7 cm/s. This assumption is a reasonable one since all f_T's were measured under a similar bias condition. Jessen[10] et al. analyzed a variety of AlGaN/GaN HEMTs without back barriers and proposed a minimal aspect ratio L_g/t_{bar} of 15 to mitigate the short channel effects, with L_g being the metallurgical gate length and t_{bar} the barrier thickness. For the devices investigated in this study, this aspect ratio value ranges from 27 to 222. Since it is much larger than 15, short channel effects are not considered here. The relatively low average electron velocity stems from the same roots why it is low in majority of the GaN-based HFETs. The high ohmic contact resistances in these devices substantially lowered the current gain, resulting in low cutoff frequencies. It has been long argued that the optical phonon emission rate is very high in GaN, e.g. ~ 10X higher than that in GaAs, while the optical phonon decay time or lifetime is comparable in GaN and GaAs.[11] As a result, it is very difficult to achieve the velocity overshoot in GaN that is commonly observed in GaAs or InP based devices. For instance, the simulation results have indicated that a gate length of ~ 25 nm is necessary to obtain velocity overshoot in GaN.[12] In order to take the full advantage of the ultrathin AlN barrier HFETs, it is necessary to fabricate devices with gate length < 50 nm with access resistance < 0.5 ohm-mm. Furthermore, a recent study[13] on SiN_x gate dielectric in AlGaN/GaN HFETs showed that SiN_x deposited by catalytic CVD resulted in the lowest gate leakage in comparison to that deposited by metalorganic CVD as well as PECVD, thanks for the low substrate temperature with no plasma employed in catalytic CVD technique; and that a thin layer of SiN_x deposited by all techniques show the similar amount of surface barrier lowering. Atomic layer deposition HfO_2 has also been reported to be promising as gate dielectric in GaN-based FETs. The future ultrascaled FETs critically depend on the development of robust gate stack, short-channel effect minimization and surface passivation schemes, so do GaN-based ones.

4. Conclusion

HFETs based on undoped AlN/GaN heterostructures with a 4 nm thin AlN barrier were fabricated using a 5 nm SiN_x deposited by PECVD as gate dielectric. The extraordinarily high concentration and mobility of the 2-dimensional electron gas led to record low sheet resistances. As a result, output current densities of 1.7 A/mm and 2.1 A/mm, intrinsic transconductance of 455 mS/mm and 785 mS/mm, were achieved for 2 μm and 250 nm gate-length devices, respectively. Current gain cut-off frequency f_T of 3.5 GHz and 60 GHz were measured on 2 μm and 250 nm gate-length devices, limited by the high ohmic contact resistance as well as the relatively long gate length in comparison to the electron mean free path under high electric fields.

Acknowledgment

The authors are thankful for discussions with Patrick Fay and Tomas Palacios and the financial support from Mark Rosker (DARPA) and Paul Maki (ONR).

References

[1] H. Xing, T. Zimmermann, D. Deen, Y. Cao, D. Jena and P. Fay. *"Ultrathin AlN/GaN heterostructure based HEMTs"*. International Conference on Compound Semiconductor Manufacturing Technology (CS ManTech), Chicago, (April 2008)

[2] H. Xing, D. Deen, Y. Cao, T. Zimmermann, P. Fay and D. Jena. *"MBE-grown ultra-shallow AlN/GaN HFET technology"*. ECS Transactions, Vol. 11, 2007.

[3] David Deen, Tom Zimmermann, Yu Cao, Debdeep Jena and Huili Grace Xing. *"2.3 nm AlN/GaN high electron mobility transistors with insulated gates"*. Phys. Solid. Stat. (c), 5(6), 2047 (2008)

[4] Tom Zimmermann, David Deen, Yu Cao, Debdeep Jena and Huili Grace Xing. *"Formation of ohmic contacts to ultra-thin channel AlN/GaN HEMTs"*. Phys. Solid. Stat. (c), 5(6), 2030 (2008)

[5] C. Wood and D. Jena. Polarization Effects in Semiconductors: From Ab Initio Theory to Device Application. Berlin, Germany: Springer-Verlag, 2007, p.522.

[6] Yu Cao, Alexei Orlov, Huili Grace Xing, and Debdeep Jena. "Very low sheet resistance and Shubnikov-de-Haas oscillations in two dimensional electron gas at ultrathin binary AlN/GaN heterojunctions". Appl. Phys. Lett. 92(15), 152112 (2008)

[7] Tom Zimmermann, David Deen, Yu Cao, John Simon, Patrick Fay, Debdeep Jena and Huili Grace Xing. *"AlN/GaN insulated gate HEMTs with 2.3 A/mm output current and 480 mS/mm transconductance"*. IEEE EDL 29(7), 661, 2008.

[8] Yu Cao and Debdeep Jena. "High-mobility window for two-dimensional electron gases at ultrathin AlN/GaN heterojunctions". Appl. Phys. Lett., 90(18), 182112 (2007).

[9] Tom Zimmermann, David Deen, Yu Cao, Debdeep Jena and Huili Grace Xing. *"Ohmic contacts to thin AlN barrier all-binary heterostructures"*. Submitted to J. Electron. Mat., 2008

[10] Gregg H. Jessen, Robert C. Fitch Jr., James K. Gillespie, Glen Via, Antonio Crespo, Derrick Langley, Daniel J. Denninghoff, Manuel Trejo Jr. and Eric R. Heller. *"Short-channel effect limitations on high-frequency operation of AlGaN/GaN HEMTs for T-gate devices"*. IEEE Trans. Electr. Dev. 2589, 2007.

[11] Kejia Wang, John Simon, Niti Goel and Debdeep Jena. *"Optical study of hot-electron transport in GaN: Signatures of the hot-phonon effect"*. Appl. Phys. Lett., 88, 022103, 2006.

[12] B. E. Foutz, S. K. O'Leary, M. S. Shur and L. F. Eastman. *"Transient electron transport in wurtzite GaN, InN and AlN"*. J. Appl. Phys., 85, 7727 (1999).

[13] Masataka Higashiwaki, Zhen Chen, Yi Pei, Rongming Chu, Stacia Keller, Nobumitsu Hirose, Takashi Mimura, Toshiaki Matsui, and Umesh K. Mishra. *"A comparative study of SiN deposition methods for millimeter-wave AlGaN/GaN HFETs"*, IEEE Device Research Conference technical digest, p.207, Santa Barbara, June 2008

International Journal of High Speed Electronics and Systems
Vol. 19, No. 1 (2009) 161–166
© World Scientific Publishing Company

EFFECT OF GATE OXIDE PROCESSES ON 4H-SiC MOSFETs ON (000-1) ORIENTED SUBSTRATE

HARSH NAIK[†], KE TANG, and T. PAUL CHOW

Center for Integrated Electronics, Rensselaer Polytechnic Institute
Troy, NY-12180 USA
[†]*naikh@rpi.edu*

We have fabricated, characterized and compared the performance of lateral n-channel 4H-SiC MOSFETs on (000-1) oriented substrates, using two different gate oxide processes. These processes include low-temperature deposited oxide and plasma-enhanced CVD oxide. Different MOSFET parameters, such as field-effect mobility, threshold voltage, Hall mobility and inversion sheet carrier concentration has been compared for the two processes.

Keywords: 4H-SiC; (000-1) SiC substrate; MOSFET; Gated Diode; Gate oxide processes

1. Introduction

SiC is a promising material for high power and high temperature electronics due to its wide energy band gap and large critical electric field [1]. SiC devices are therefore considered as an alternative to silicon devices for power electronics applications especially under extreme environments. High-voltage 4H-SiC MOSFETs have been preferred over 6H-SiC because 4H-SiC polytype has a higher and more isotropic bulk mobility when compared to 6H-SiC. However, one of key obstacles in 4H-SiC power MOS devices, such as the power MOSFET and IGBT, is the inferior electrical characteristics of SiO_2/4H-SiC interface, when compared to Si MOS, mainly due to the presence of large acceptor-like interface states near the conduction band edge which leads to a low inversion layer mobility. Over the years, different research groups have tried different gate dielectrics, annealing conditions and substrates to improve the inversion layer mobility [2-5].

In this paper, we compare the performance of 4H-SiC MOSFETs with two different gate oxide processes on (000-1) or carbon-face SiC substrates.

2. Device Structures and Fabrication

A schematic cross-section of the fabricated SiC MOSFET is shown in Fig. 1 which was used to characterize the two different gate oxide processes.

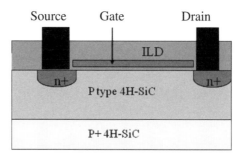

Fig. 1 Schematic cross-section of a lateral n-channel SiC MOSFETs

Two different sets of MOSFETs were fabricated, with 8°-off (000-1) substrates with p-type epi layer doped to $3.5 \times 10^{16} cm^{-3}$ on p+ substrates. Source and drain regions were selectively implanted with a phosphorous dose of 4×10^{15} cm^{-2} and a maximum energy of 195 keV. The wafers were then cleaned with RCA clean and diluted HCl before 1um-thick field oxide was deposited. Annealing at 1600°C for 10 min. in Ar was performed to activate the implanted phosphorous. Graphite capping layer was used to prevent dopant out-diffusion during implant activation. One set of MOSFETs has Plasma-Enhanced CVD TEOS (PTEOS) of 100nm deposited in a low-pressure plasma reactor at 400°C, followed by re-oxidation in NO at 1175°C for 2 hr. Another set has a gate oxide of 100nm-thick low-temperature oxide (LTO), followed by a re-oxidation in NO at 1175°C for 2 hrs. Subsequently, a 650 nm thick polysilicon was deposited and degenerately doped using POCl₃. After definition of the gate, 1μm thick SiO₂ was deposited to serve as inter-level dielectric. The source and drain contacts (Ti/Ni/Al) were defined using liftoff technique, and the ohmic contacts were annealed at 1000°C for 2min. in Ar ambient. The gate contact was patterned and etched, followed by (Ti/Mo) contact metallization.

3. Experimental Results

3.1. *MOS-Gated Diode Measurements*

The C-V measurements were done on the gated diode devices fabricated along with the MOSFETs. Fig. 2 shows the results of the C-V measurements compared with the C-V curve for the ideal case. As seen from the figure these devices exhibit a 'hook and ledge' feature observed in 4H-SiC and 6H-SiC [5,6] before due to the presence of large number of surface states in the bandgap. The interface state charges can be estimated from the 'hook' feature in the gated diode curve from accumulation to depletion, using the formula:

$$N_{it} = \frac{\Delta V C_{ox}}{q} \qquad (1)$$

where ΔV is the 'hook' voltage. This value of N_{it} was calculated as 1e13cm^{-2} and 1.1e13cm^{-2} for PTEOS and LTO samples respectively.

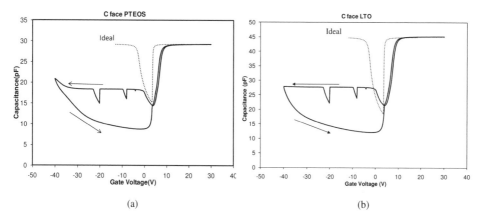

Fig.2 Gated diode characteristics of C face (a) PTEOS and (b) LTO samples.

3.2.   *MOSFET Measurements*

Fig. 3 shows the transfer characteristics and the field-effect mobility for the two different processes. Due to the 'soft turn-on' of SiC MOSFETs we need to define a low-current and a high-current threshold voltage parameter for the SiC MOSFETs. The low-current threshold is determined from the current, a certain fraction of the turn-on current, at which the MOSFET turns on. The high-current threshold is defined by the extrapolation of the linear region of the transfer I_D - V_G curve.

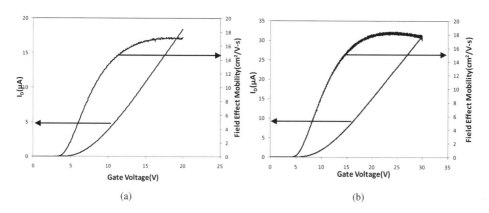

Fig.3 Transfer characteristics and field-effect mobility of SiC MOSFETs with (a) PTEOS and, (b) LTO gate oxide process

The low-current and high-current threshold for PTEOS sample are 5V and 6.5V and for the LTO sample the low and high current threshold voltages are 6V and 8V, respectively. High threshold voltage values are due to the presence of large interfacial charges, seen from the MOS-gated diode measurements. Larger interfacial charges for the LTO process leads to a larger threshold voltage for LTO sample. This also suggests that the sub-threshold should be better for the PTEOS sample compared to the LTO sample, which is what was observed, with a sub-threshold slope of 400mV/dec and 600mV/dec for the PTEOS and LTO sample respectively. This sub-threshold is better than 1000mV/dec and 700mV/dec for the PTEOS and LTO samples on the (0001) or Si face substrates fabricated at the same time, thus indicating a lower density of interface states near the conduction band edge for carbon-face samples when compared to silicon-face ones.

Fig. 3 also shows the field-effect mobility of the two samples. The field-effect mobility is similar for both the samples, with peak mobility of about 19cm^2/V-s for LTO, which slightly higher than the 17cm^2/V-s for PTEOS.

3.3. *MOS-Gated Hall Bar Measurements*

Hall characterization was also done on the MOS-gated Hall bars fabricated at the same time with the MOSFETs. Using the Hall measurements the inversion sheet carrier concentration and the intrinsic MOS electron mobility can be determined independent of each other. In Figure 4 Hall mobility is plotted as a function of effective surface field, which is calculated using the formula shown here [7]:

$$E_{eff} = \frac{1}{\varepsilon_s}(Q_b + \frac{Q_n}{2}) \qquad (2)$$

and hence,

$$E_{eff} = \frac{1}{\varepsilon_s}(sqrt(2\varepsilon_s N_A(2\phi_b + V_{SB}) + \frac{qN_s}{2}) \qquad (3)$$

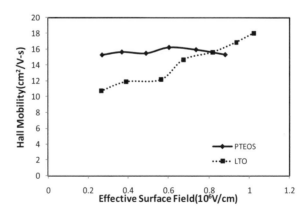

Fig.4 Hall mobility plotted as a function of effective surface field

Thus, it can be seen that both samples also have similar Hall mobilities. In Fig. 5 the inversion sheet carrier concentration is plotted and compared with the ideal sheet carrier concentration from the Charge Sheet Model (CSM). Also plotted is the sheet carrier concentration using an approximation of the CSM:

$$qN_s = C_{ox}(V_G - V_T) \qquad (4)$$

where V_T is calculated form the extrapolation of the N_s vs. V_G curve. The slope of the initial points at low gate voltages is different form high gate voltages, this is due to the 'soft turn on' of SiC MOSFET, and hence V_T was extracted from the extrapolation of first two points. The difference between the ideal CSM and this approximate formula will be due the total effective oxide charges ($Q_f + Q_{it}$) and the difference between the measured and the approximate formula will be due to the interface trapped charges.

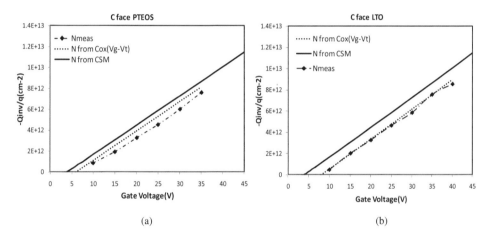

Fig.5 Inversion Sheet Carrier Concentration compared with the ideal Charge Sheet Model

It can be observed that the LTO sample has more oxide charges compared to the PTEOS sample. By contrast, there is no apparent carrier trapping for LTO sample and $\sim 10^{12}/cm^2$ for the PTEOS sample. These interface charges seen are from the strong inversion point in the bandgap to the conduction band edge ($\sim 0.2eV$). These low interfacial charges should not be confused with the large interface charges seen in MOS-gated diode measurements, since the gated diode measurement scans across a larger range in the bandgap from accumulation to strong inversion ($\sim 2.5eV$) and hence it shows larger interface charges than seen from the MOS-gated Hall bar measurements. The interface states seen in MOS-gate diode measurement from the weak inversion to the strong inversion and are in the upper half of the bandgap and will be filled with electrons, hence acting as scattering centers and will degrade the mobility. Also, it can be seen from the MOS-gated diode measurements that measured curve is more close to the ideal curve near inversion than in accumulation, which corresponds to the fact that there is less carrier trapping near the conduction band edge.

4. Summary

We have performed electrical characterization of MOS-gated diodes, MOSFETs and MOS-gated Hall bars to compare two different gate oxide processes on 4H-SiC MOSFETs on (000-1) C-face substrates. It was found that the both processes give similar performances of the MOSFETs in terms of field-effect mobility and Hall mobility. Also carrier trapping near the conduction band edge was found to be very small, with the LTO process giving a marginally better performance. High interfacial charges were observed from weak inversion to strong inversion from the gated diode measurements.

Acknowledgement

This work was supported by NSF Center for Power Electronic Systems (# EEC-9731677) and GE Global Research Center.

References

1. T. P. Chow and R. Tyagi, "Wide bandgap compound semiconductors for superior high- voltage unipolar power devices," *IEEE Trans. Electron Devices*, vol. 41, no. 8, pp. 1481-1483, Aug. 1994.
2. G.Y. Chung, C.C. Tin, J.R. Williams, K. McDonald, M. Di Ventra, S.T. Pantelides, L.C. Feldman, and R.A. Weller, "Effect of nitric oxide annealing on the interface trap densities near the band edges in the 4H polytype of silicon carbide," *Appl. Phys. Lett.*, vol. 76, no.13, pp. 1713-1715, Mar. 2000.
3. G. Gudjónsson, H. Ö. Ólafsson, and E. Ö. Sveinbjörnsson, "Enhancement of inversion channel mobility in 4H-SiC MOSFETs using a gate oxide grown in nitrous oxide (N_2O)," *Mater. Sci. Forum*, vol. 457-460, pp. 1425-1428, 2004.
4. M. Okamoto, S. Suzuki, M. Kato, T. Yatsuo, and K. Fukuda, "Lateral RESURF MOSFET Fabricated on 4H-SiC (000-1) C-Face," *IEEE Electron Device Lett.,* vol. 25, no.6, pp. 405-407, Jun. 2004.
5. Y. Wang, K. Tang, T. Khan, M.K. Balasubramanian, H. Naik, W. Wang, T.P. Chow, "The Effect of Gate Oxide Processes on the Performance of 4H-SiC MOSFETs and Gate-Controlled Diodes," *IEEE Trans. on Electron Devices*, 2008 *(in press)*.
6. S. T. Sheppard, J. A. Cooper, and M. A. Melloch, "Nonequilibrium characteristics of the gate-controlled diode in 6H-SiC," *J. Appl. Phys.*, vol. 75, no.6, pp. 3205–3207, Mar. 1994.
7. A. G. Sabnis and J. T. Clemens, "Characterization of the electron mobility in the inverted 100 silicon surface," in *IEDM Tech. Dig.*, 1979, pp. 18–21.

International Journal of High Speed Electronics and Systems
Vol. 19, No. 1 (2009) 167–172
© World Scientific Publishing Company

CHARACTERIZATION AND MODELING OF INTEGRATED DIODE IN 1.2kV 4H-SiC VERTICAL POWER MOSFET

HARSH NAIK[†], YI WANG, and T. PAUL CHOW

Center for Integrated Electronics, Rensselaer Polytechnic Institute
Troy, NY-12180 USA
[†]naikh@rpi.edu

We have characterized and modeled the integrated diode of a 1.2kV 4H-SiC power MOSFET. We have measured its static and dynamic characteristics up to 200°C and extracted relevant SPICE model parameters. From the extracted turn-on voltage and ideality factors, we conclude that the integral diode is not a pin junction diode, but a unipolar diode.

Keywords: 4H-SiC; Vertical MOSFET; Integrated diode

1. Introduction

SiC is a promising material for high power and high temperature electronics due to its wide energy band gap and large critical electric field [1]. Based on intrinsic material properties, SiC high-voltage power semiconductor devices have been projected to have much better (> 100 times lower specific on-resistance) performance than equivalent silicon devices of the same voltage ratings. Many high-voltage experimental devices have been demonstrated, with increasing blocking voltages, up to 20kV in power rectifiers and 10kV in power transistors. In addition, commercial 4H-SiC Schottky rectifiers (diodes) up to 1200V, 20A ratings as well as prototype JFETs and MOSFETs up to 1200V and 5A are now available As two of voltage-controlled three-terminal switches, JFETs and MOSFETs could be used in motor drive circuits, in which a switch with a current-limiting feature is desirable for over-current protection.

As a voltage-controlled three-terminal switch, the MOSFET can be used in motor drive circuits, in which a switch with a current-limiting feature is desirable for over-current protection. Some circuits however require reverse conduction through the active power-switching device. When using a bipolar transistor as a switch, an anti-parallel diode is used to facilitate reverse conduction. When power MOSFET is used as a switch, the inherent diode in the MOSFET structure can be used for reverse conduction, thus eliminating the complexity of the circuit without an additional cost of an external diode. The presence of this diode can however, restrict the efficiency and speed of the circuit due to excess stored charge in the drift region which forces a high reverse current during reverse recovery. This also limits the safe-operating-area (SOA) and the switching speed of the device. To overcome this, a Junction-Barrier Schottky (JBS) diode can be integrated in the same die in parallel with the p-body/n drift junction diode, without affecting the normal operation of the MOSFET [2, 3].

In this paper, we characterized and modeled the integrated diode of a 1.2kV 4H-SiC power MOSFET from Cree Inc.

2.   Characterization of the Integral Diode

A schematic cross-section of a vertical high-voltage SiC MOSFET is shown in Fig. 1. The integrated diode can be either pin junction diode type or it can have Schottky region in between the implanted p-regions i.e. JBS type [3], to enhance its performance.

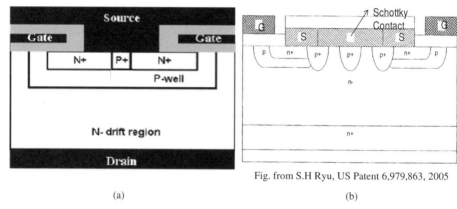

Fig. from S.H Ryu, US Patent 6,979,863, 2005

(a)   (b)

Fig.1 Schematic cross-section of a vertical power MOSFET with an integrated (a) pin type (b) JBS type diode.

2.1.   *Forward I-V Characteristics*

The MOSFET chips were mounted on high-temperature packages which can withstand up to at least 350°C. The static and dynamic characteristics of the MOSFET have been measured up to 200°C and modeled and compared to those of commercial SiC JFET [4]. Fig. 2 shows the forward I-V characteristics of the integrated diode. I-V characteristics were measured up to 200°C and the forward drop of the diode decreases with temperature due to the decrease in built-in potential with temperature.

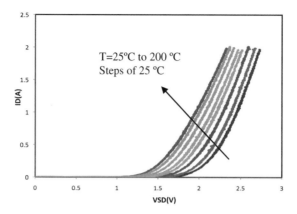

Fig.2 Forward I-V characteristics of the diode up to 200°C.

In the forward I-V characteristics, we have found three distinct regimes, namely, the very low-current level region, the low-current ideal-diode exponential region and the high-current level region. The low-current region, i.e. the thermionic emission region is shown in Fig. 3.

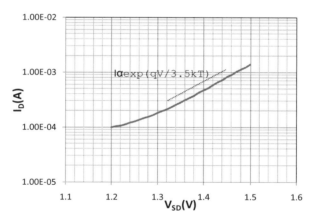

Fig. 3 Forward I-V of the diode in the low-current region in a semi-log scale.

In this regime the forward voltage is between 1.1 and 1.5 V, and the ideal diode equation was found to hold but the ideality factor extracted is rather large (3.5). This extracted 'n' value is greater than 1 indicating a deviation from the thermionic emission theory. The turn-on at such low voltages points to a unipolar type of integrated diode. The forward I-V for the very low-current level and high-current level case is plotted in Fig. 4. Due to some damage at the metal/SiC junction, charge-trapping centers are introduced near the junction; hence the current flow is limited by space charges and is characterized by power law [5], instead of exponential dependence. Thus, we see power-law dependence in the very low-current level and the high-current level case.

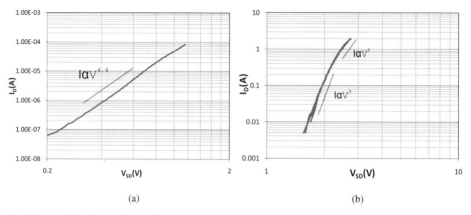

(a)   (b)

Fig. 4 Forward I-V characteristics of the integrated diode plotted in log –log scale for (a) very low-current level, (b) high-current level.

Thus, in our case there are actually three regions. For $V_F < 1.1V$, at very low-current level, the I-V characteristics in this space charge limited region is best shown as $I \propto V^n$ where n has been found to be 4.4.   Second region is the exponential region described earlier. And thirdly, at high-current level ($I_F > 1A$), the power law again applies and the 'n' factor has been found to be 9 for $V_F = 1.5$ to 2.5V and 6 for $V_F > 2.5V$. The current is series resistance limited in this region and some minority carrier can also injected due to increase in drift component of the current at high fields.

2.2. *Reverse I-V Characteristics*

The reverse leakage current of the diode was measured up to 200°C. It is plotted in Fig. 5.

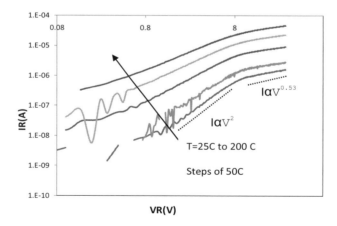

Fig.5 Reverse I-V characteristics of the integrated diode on log-log scale.

There are two distinct regions in the reverse characteristics. It can be seen that, initially, the reverse current rises faster, with a power law and an 'n' value of 2, this is the space charge limited current. After V_R of about 8V the reverse current increases with a power law exponent of 0.53, which indicates a space charge generation limited junction leakage current.

2.3. *Reverse Recovery Characteristics*

Reverse recovery measurements were done up to 200°C and the result is plotted in Fig. 6.

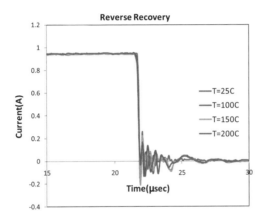

Fig.6 Reverse Recovery characteristics of the integrated diode.

Since the reverse recovery is very fast which does not increase with temperature and shows little stored charge, it further corroborates to our assertion that the integrated diode is not a pin junction type, but a unipolar type device. We get a very small reverse recovery current; the small reverse current seen could be just due to some stray capacitances in the circuit.

3. Modeling of the Integrated Diode

Finally, the SPICE parameters for the diode were extracted from the forward characteristics and the temperature characterization of the integral diode. The saturation current (IS) was extracted from the current intercept the diode I-V curve at zero voltage bias. The diode resistance (RS) was extracted from the slope of the I-V curve. The ideality factor (N) was extracted as 3.5 using the methods described in Section 2.1. The temperature coefficient (XTI) was calculated by assuming the power law dependence of the reverse current on temperature. The extracted parameters are shown in the table in Fig. 7.

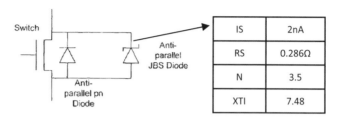

	IS	2nA
	RS	0.286Ω
	N	3.5
	XTI	7.48

Fig.7 Extracted SPICE parameters of the integrated diode.

4. Conclusion

We have characterized the integrated diode of a 1.2kV 4H-SiC power MOSFET from Cree and have found three distinct regions of operation in forward conduction. The very low-current level region is space charge limited and the high-current level region is series resistance limited, hence exhibit a power law rather than exponential. From the forward and reverse I-V and reverse recovery characteristics it was inferred that the integrated diode is a unipolar device but it is not a Schottky barrier diode, since this diode has a turn on voltage of ~2V compared to a turn on voltage of ~1V for commercial SiC Schottky rectifiers. Also from the temperature dependence of the forward I-V curves we can see that the forward voltage drop decreases with increasing temperature, unlike Schottky diodes for which the forward drop initially decreases and then starts increasing with temperature at higher current levels due to series resistance limited current. Thus, we conclude that the diode is a MOS Controlled Diode [6]. The SPICE model parameters of the integrated diode are also extracted.

Acknowledgement

This work was supported by Boeing and NSF Center for Power Electronic Systems (# EEC-9731677).

References

1. T. P. Chow and R. Tyagi, "Wide bandgap compound semiconductors for superior high-voltage unipolar power devices," *IEEE Trans. Electron Devices*, vol. 41, no. 8, pp 1481-1483, Aug. 1994.
2. K. Mondal R. Natarajan and T.P. Chow, "An Integrated 500-V Power DMOSFET/Antiparallel Rectifier Device With Improved Diode Reverse Recovery Characteristics," *Electron Device Lett.,* vol. 23, no. 9, pp. 562-564, Sept. 2002.
3. S.-H. Ryu, "Silicon Carbide MOSFET with Integrated Anti-Parallel Junction Barrier Schottky Freewheeling Diodes and Methods of Fabricating the Same", US Patent 6,979,863, 2005.
4. Y. Wang et. al., "Modeling of High Voltage 4H-SiC JFETs and MOSFETs for Power Electronics Applications", *Power Electronics Specialist Conference (PESC),* 2008.
5. S.K. Ghandhi, *Semiconductor Power Devices*, John Wiley & Sons, New York, 1977.
6. Q. Huang and G.A.J Amaratunga, "MOS Controlled Diodes-A New Power Diode", Solid State Electronics, vol. 38, no.5, pp. 977-980, 1995.

International Journal of High Speed Electronics and Systems
Vol. 19, No. 1 (2009) 173–181
© World Scientific Publishing Company

PACKAGING AND WIDE-PULSE SWITCHING OF 4 MM x 4 MM SILICON CARBIDE GTOs

HEATHER O'BRIEN

U.S. Army Research Laboratory, 2800 Powder Mill Road,
Adelphi, MD, USA
hobrien@arl.army.mil

M. GAIL KOEBKE

U.S. Army Research Laboratory, 2800 Powder Mill Road,
Adelphi, MD, USA
gkoebke@arl.army.mil

The U. S. Army Research Laboratory (ARL) is investigating compact, energy-dense electronic components to realize high-power, vehicle-mounted survivability and lethality systems. These applications require switching components that are low in weight and volume, exhibit reliable performance, and are easy to integrate into the vehicles' systems. The devices reported here are 4 mm x 4 mm silicon carbide GTOs rated for 3000 V blocking. These devices were packaged at ARL for high pulse current capability, high voltage protection, and minimum package inductance. The GTOs were switched in a 1-ms half-sine, single-pulse discharge circuit to determine reliable peak current and recovery time (or T_q). The GTOs were repeatedly switched over 300 A peak (3.3 A/cm^2 and an action of 60 A^{2}s) with a recovery time of 20 μs. The switches were also evaluated for dV/dt immunity up to an instantaneous slope of 3 kV/μs.

Keywords: silicon carbide; pulsed power

1. Introduction

Future Army applications will require several mega-joules of energy to be generated, stored, and released while staying within the tight volumetric and weight confines of current vehicles. This necessitates switching components that are lighter weight and twice the power density of the present technology. Silicon carbide is the only material with the potential to meet these requirements within the next ten years. Earlier SiC thyristors and GTOs have shown great potential in pulse switching applications [1, 2]. While material growth and device technology continue to improve, the future switching requirements need to be accurately modeled in the laboratory in order to assess the switches at appropriate peak currents, pulse widths, and recovery times. To achieve the highest performance that advanced SiC devices can provide, new packaging technologies are required.

ARL has been evaluating both silicon carbide and silicon Super-GTO (or SGTO) chips at slow-rising currents up through the kilo-amp range and at pulse widths up to one millisecond. Once understanding the full capabilities of the individual die, optimal driving circuits and current bussing can be determined for scaled-up integrated switches of parallel- and series-connected devices that would meet the requirements of high-current, high-action (I^2t) applications. The SGTOs have been designed and tested for a wide range of Army applications with different current, rise time, and recovery requirements. The present test beds in use model an application that calls for high-action pulse capabilities with simple controls in a power-dense package. Silicon-based SGTOs are being considered as a near-term switching solution for system demonstrations, while silicon carbide is being explored as a future replacement for the silicon once SiC growth and processing matures. The silicon carbide SGTOs detailed in this paper were designed and fabricated through cooperative agreements with Silicon Power Corporation and Cree, Inc., then packaged and pulse tested at ARL for comparison to both the silicon die and larger area thyristors [3]. This paper describes the silicon carbide packaging, test beds, and device performance.

2. Device and Packaging

The silicon carbide SGTO has a footprint of 0.16 cm^2 and a central active area measuring 0.09 cm^2, excluding the edge termination (Fig. 1). Typical high voltage hold-off for this asymmetric switch is 3000 V. For wide-pulse switching, the gate is triggered negative relative to the anode with a pulse width of 30 μs. These SGTOs have finer anode/gate structure and higher voltage hold-off than previously studied thyristors from Cree, but perform similarly in high current pulse circuits [4]. Devices were statically characterized on a curve tracer prior to being inserted into the pulse test bed. The typical I-V curves for forward blocking and turn-on are shown in Figs. 2 and 3.

Fig. 1. Anode/gate surface of the 4 mm x 4 mm SiC SGTO.

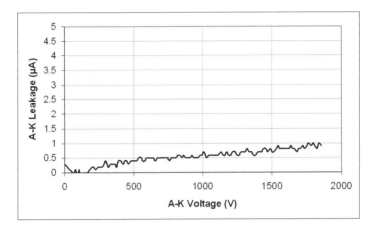

Fig. 2. Typical low anode-cathode current leakage for the SGTO.

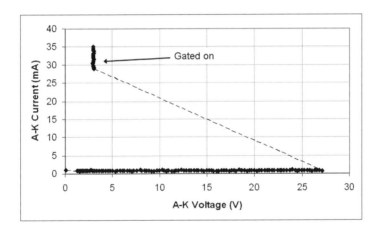

Fig. 3. Typical voltage drop and anode-cathode holding current for the SGTO.

Various types of SGTO packaging were developed based on careful analysis of the device and on ARL's experience with packaging for high-power, pulsed applications. The SGTO package primarily used for these evaluations is prepared by mounting each chip on a direct bond copper substrate using eutectic AuSn die attach. The small footprint of the devices and the fine separation between anode and gate areas necessitates several one-mil wire bonds to make the direct connections from the anode pads of the device to a copper bus suitable for linking to the pulse circuit. The small copper bus bars connected to the cathode base and the anode are laminated with a ceramic dielectric and coated with high-voltage thermo plastic to isolate the tabs from each other while still maintaining low package inductance. The three center gate pads are bonded to an intermediate insulated block on the opposite side of the package using one-mil wires. Ten-mil wire bonds are used to connect this intermediate block to the copper gate extension on the package

(Fig. 4). To ensure proper operation and prevent flashover at high voltage, the packages are encased in a high-temperature epoxy that was custom-developed to allow for flexing of the wires during pulsing while maintaining wire spacing. The downside of using this epoxy is that it is difficult to see through and to remove for analysis of failed devices.

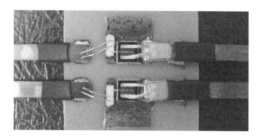

Fig. 4. Wire-bonded SGTOs prior to encapsulation. The SGTOs were mounted on the same board but operated individually. Separate gate tabs for each extend to the left; anode and cathode tabs extend to the right.

More recently, a commercially-available power package has been employed to investigate the use of five-mil wire bonds and a new clear encapsulation which would simplify the packaging and failure analysis processes. The five-mil wire is large enough to significantly reduce the number of bonds required for the high pulse current, but also small enough to avoid damage to the thin gold metallization on the anode and gate of the GTO.

3. Pulse Evaluation

A simulation program was used to design a high-energy test circuit that could provide a current pulse with a half-sine shape and a peak current up to 1000 A from an initial charge well within the voltage limitations of the SGTOs. A relatively large series inductor was used to spread the energy discharge over time. Diodes clamped the ringing that resulted from the under-damped condition. A schematic of the final circuit design and a screen capture of the simulated waveforms are shown in Fig. 5. This circuit produced a current pulse with a half-sine shape and a base pulse width of 1 ms (or 500 Hz). The expected peak current was around 800 A, based on preliminary silicon devices' performance and on earlier narrow pulse testing of both types of SGTO [5, 6].

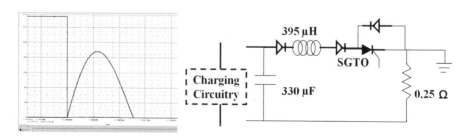

Fig. 5. Simulation of an RLC circuit designed to provide a half-sine shaped current pulse.

Recovery time, or T_q, evaluations were also done with the SiC SGTOs. The recovery time was defined as the delay time between when the initial pulse current fell to zero and when the GTO's anode could block high voltage again. This device parameter is important in applications that have multiple coordinated switching events. A hard turn-off of cathode current is not required in the present application, so the full turn-off capability of the SGTOs was not explored at this time. The same 1-ms test bed was used, and a separate power supply, small capacitor bank, and IGBT were used to reapply the high voltage. Additional diodes were added to the circuit so that the main capacitor bank was not being recharged. An adjustable R-C shaping network set the rise time of the reapplied voltage. A general block diagram of the T_q test set-up is shown in Fig. 6. No assisted gate turn-off was required for the SiC devices, simplifying the gate drive controls. During the T_q evaluations, the SGTOs were turned on with the same basic gate driver used when evaluating the switches for peak pulse current. It was expected that the silicon carbide devices would recover more rapidly than similar 3.5 cm^2 silicon devices based on material properties and literature research [7, 8].

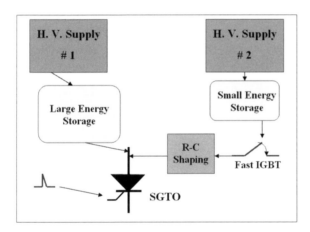

Fig. 6. Generalized schematic showing how the recovery circuit (r.) was added to the main pulse circuit (l.).

Parts of the T_q circuitry were also used to evaluate the dV/dt immunity of the SGTOs. It is desirable for the switches to remain in the off-state in noisy environments until they are actively triggered on. The resistance and capacitance of the T_q circuit were dropped in order to more rapidly apply voltage at the anode of the SGTO. It was found that adding a small amount of capacitance directly at the anode and the gate of the switch package allowed the devices to withstand much higher dV/dt without turning on (Fig 7). This set of tests was conducted without high-current pulsing of the switch; it was only intended to determine noise immunity.

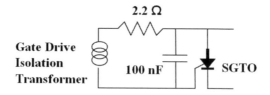

Fig. 7. Gate drive connection to the SGTO for dV/dt immunity testing.

4. Results

For the 1-ms current pulse, 300-320 A was determined to be the reliable, repeatable peak current for these 0.16 cm^2 SGTOs (Fig. 8). This corresponds to a current density of 3.3 kA/cm^2 over the central mesa area of the device. The action level was calculated to be 60 A^{2}s. As mentioned, the pulse circuit was designed to operate at higher voltage and current levels, because it was predicted that these SGTOs would switch at a much higher action level compared to the level calculated from earlier narrow-pulse switching [4]. The SiC devices are still switching at a higher current density than the silicon SGTOs (which pulsed at 2.5 kA/cm^2), though a better comparison will be made in the future when larger SiC chips become available.

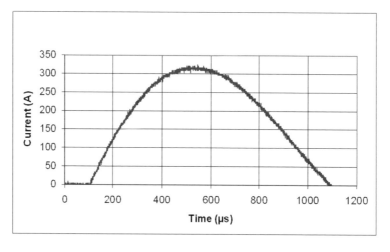

Fig. 8. Peak pulse current of 320 A.

Devices that were pulsed several times beyond 320 A tended to fail short in the anode-to-gate and were no longer able to be triggered on. It is hypothesized that the passivation at the gate is sustaining damage due to the temperature rise over the course of the wide anode-cathode pulse current. As previously stated, the hard, caramel-colored epoxy used to protect the device makes post-failure visual analysis difficult. Now that different packaging has been developed with fewer wire bonds and a clear low-viscosity silicone elastomer, microscopic and SEM images can be collected to better view the surface of a failed SGTO. These results will be reported at a later date.

Recovery time for the SGTOs was measured following a 300 A current pulse. The shortest T_q measured was 20 μs (Fig 9). The reapplied voltage signal had an instantaneous dV/dt of 30 V/μs which was similar to the slope of the falling current. No turn-off signal was sent to the gate in order to obtain this recovery time. The T_q depends on the charge carrier distribution at the high voltage p-n junction and the anode-gate p-n junction at the time that voltage is reapplied at the anode [9]. The recovery measurement can also be affected by the SiC's temperature rise over the course of the initial switching event, which would have been higher had the SGTO been able to handle a higher pulse current.

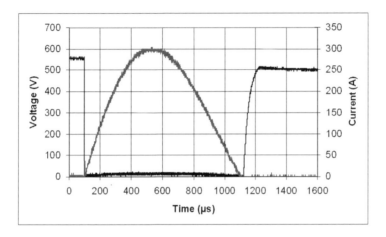

Fig. 9. Forward recovery time of SiC SGTO with device blocking voltage after 20 μs.

Without any gate-triggering or high current pulsing, the SGTOs withstood a voltage transient of about 3 kV/μs peak slope (Fig 10). Low-voltage capacitors totaling 200 nF were added at the gate in order to redirect anode-gate current flow generated by this dV/dt. Experiments were also conducted with a single 100 nF, 20 V capacitor at the gate, but dV/dt immunity in that case was unreliable, so 200 nF was chosen as the ideal capacitance. In this configuration, the SGTO could still be triggered on with low gate current, but was also capable of reasonable noise immunity relative to its high voltage hold-off rating. The results relieved concerns that the silicon carbide switches may be much more susceptible to dV/dt turn-on than silicon switches because of the gate triggering relative to the anode where the high voltage was applied.

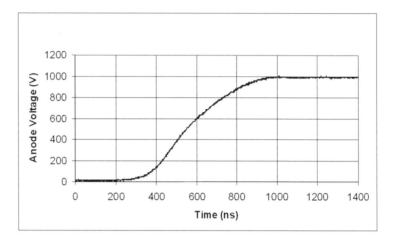

Fig. 10. Demonstrated dV/dt immunity of 3 kV/μs at the fastest portion of the slope. Note time scale.

5. Conclusion

The 0.16 cm^2 silicon carbide SGTO was evaluated in a wide-pulse circuit up to 320 A, 3.3 kA/cm^2 over the mesa area and an action of 60 A^{2}s. The forward recovery time following such a high current pulse was measured to be 20 μs. The SGTO was also found to be immune to an instantaneous dV/dt of 3 kV/μs. As predicted from narrow-pulse research previously reported by ARL [7], silicon carbide was switched at higher current density and faster un-assisted recovery times than silicon. These results encourage the further development of larger area, higher voltage SiC GTOs. ARL is currently preparing to package and evaluate 0.5 cm^2 SiC SGTOs to determine whether device performance will scale.

References

[1] S. B. Bayne and D. Ibitayo, "Evaluation of SiC GTOs for pulse power switching," Proc. 14th IEEE Pulsed Power Conf., pp. 135-138, 2003.

[2] S. Ryu, A. K. Agarwal, R. Singh, and J. W. Palmour, "3100 V, asymmetrical, gate turn-off (GTO) thyristors in 4H-SiC," *IEEE Electron Device Letters*, vol. 22, issue 3, pp. 127-129, 2001.

[3] V. Temple, " 'Super' GTO's push the limits of thyristor physics," Proc. 35th Annu. IEEE Power Electronics Specialists Conf., pp. 604-610, 2004.

[4] H. O'Brien, C. J. Scozzie, S. B. Bayne, and W. Shaheen, "Overview of pulsed power research at the Army Research Laboratory," Proc. 33rd Annual GOMACTech Conference, 17-20 March 2008, Las Vegas, NV.

[5] H. O'Brien, W. Shaheen, T. Crowley, and S. B. Bayne, "Evaluation of the safe operating area of a 2.0 cm^2 Si SGTO," Proc. of the 2007 IEEE Pulsed Power and Plasma Science Conf., vol. 2, pp. 1034-1039, June 2007.

[6] H. O'Brien, W. Shaheen, R.L. Thomas, Jr., T. Crowley, S.B. Bayne, and C. J. Scozzie, "Evaluation of advanced Si and SiC switching components for Army pulsed power applications," *IEEE Trans. Magn.*, vol. 43, no. 1, pp. 259-264, Jan. 2007.

[7] "Silicon carbide substrates: product specifications," Cree, Inc., Durham, NC, Version MAT-CATALOG.00D, Revised Jan. 2005.

[8] T. Burke, K. Xie, H. Singh, T. Podlesak, J. Flemish, J. Carter, S. Schneider, and J. Zhao, "Silicon carbide thyristors for electric guns," *IEEE Trans. Magn.*, vol. 33, no. 1, pp.432-437, Jan. 1997.

[9] B. Jayant Baliga, Power Semiconductor Devices. Boston: PWS Publishing Co., 1996.

International Journal of High Speed Electronics and Systems
Vol. 19, No. 1 (2009) 183–192
© World Scientific Publishing Company

BI-DIRECTIONAL SCALABLE SOLID-STATE CIRCUIT BREAKERS FOR HYBRID-ELECTRIC VEHICLES

D. P. URCIUOLI

Sensors and Electron Devices Directorate, U. S. Army Research Laboratory, 2800 Powder Mill Road, Adelphi, MD, 20783, USA
durciuoli@arl.army.mil

VICTOR VELIADIS

Compound Semiconductor Group, Northrop Grumman Corporation, 1212 Winterson Road, Linthicum, MD, 21090, USA
victor.veliadis@ngc.com

Power electronics in hybrid-electric military ground vehicles require fast fault isolation, and benefit additionally from bi-directional fault isolation. To prevent system damage or failure, maximum fault current interrupt speeds in tens to hundreds of microseconds are necessary. While inherently providing bi-directional fault isolation, mechanical contactors and circuit breakers do not provide adequate actuation speeds, and suffer severe degradation during repeated fault isolation. Instead, it is desired to use a scalable array of solid-state devices as a solid-state circuit breaker (SSCB) having a collectively low conduction loss to provide large current handling capability and fast transition speed for current interruption. Although, both silicon-carbide (SiC) JFET and SiC MOSFET devices having high breakdown voltages and low drain-to-source resistances have been developed, neither device structure alone is capable of reverse blocking at full voltage. Limitations exist for using a dual common-source structure for either device type. Small-scale SSCB experiments were conducted using 0.03 cm^2 normally-on SiC VJFETs. Based on results of these tests, a normally-on VJFET device modification is made, and a proposed symmetric SiC JFET is considered for this application.

Keywords: SSCB; bi-directional; VJFET; SiC

1. Introduction

As the military transitions to more electric and hybrid-electric vehicle systems, high-power electronics are filling an increasing number of mission critical roles. For future hybrid-electric ground vehicles, a wide range of converters operating at up to 600 VDC and up to hundreds of kilowatts are being developed. To prevent damage to converters or other system components during fault conditions, fault current interrupt speeds in tens to hundreds of microseconds are necessary. Furthermore, inverter / rectifier systems and some DC-DC converters operating between two voltage busses having independent sourcing capability, require bi-directional fault isolation. Because mechanical contactors do not provide adequate actuation speeds, and suffer severe degradation during repeated fault isolation, a solid-state circuit breaker (SSCB) is desired. The high conduction losses and negative temperature coefficients of bipolar devices limit their practicality and scalability for this application. Instead, it is desired to use a parallel set of majority carrier devices having a collectively low conduction loss to provide large current handling capability and fast transition speed for current interruption. Recently developed 1200 V silicon-carbide (SiC) JFET and MOSFET devices have low drain-to-source resistances, making them more viable candidates for use in scalable SSCB modules.

2. Mitigation of Induced Voltage Spikes

With the increased transition speed of a SSCB module compared to that of a mechanical device, mitigation of inductive voltage spikes using snubbers must be considered. Abruptly arresting large currents flowing through parasitic inductances of power system cabling, bus bars, and interconnects can induce voltage spikes in excess of the breakdown voltage rating of most semiconductor devices. To preserve the two terminal structure of an ideal bi-directional fault protection module, which may not have either terminal referenced to ground, limits the selection of snubber topologies. Furthermore, the selected snubber topology must be compatible with bi-directional current flow through the module. Figure 1 shows the bi-directional snubber topology selected, in the context of a high-power, single-phase, DC-DC, bi-directional converter for regulating bus voltage and battery state of charge in a future combat systems hybrid-electric vehicle.

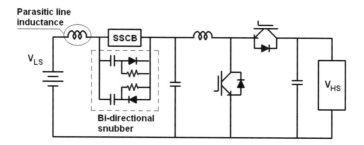

Fig. 1. Bi-directional RCD snubber and SSCB module in a DC-DC converter application.

Assuming a parasitic line inductance of ten microhenries, simulations were conducted to determine suitable snubber capacitor values to prevent overvoltage on a 1200 V rated SSCB. For example, a fault that causes current to rise through a system inductance is detected in a SSCB initially conducting 500 A, and the SSCB is actuated to a fully open state within 40 μs, with a transition time of 50 ns. The result is a peak SSCB voltage of approximately 1000 V using 1.4 μF snubber capacitors for an interrupted current of over 900 A. Low average current rated 1200 V SiC junction barrier schottky diodes having high peak current ratings can be used. A wide range of snubber resistor values are applicable depending on the frequency of SSCB actuation.

3. SiC MOSFET Approach

The U.S. Army Research Laboratory has been working with Cree, Inc. and Northrop Grumman Corporation (NGC) in the development of SiC power transistors for the past several years. Technology application discussions with Cree in late 2007 led to a feasibility study of a parallel SiC MOSFET based replacement for mechanical contactors used for fault protection in an Army hybrid-electric vehicle technology demonstrator. It was later determined that the most practical application of the module required bi-directional fault protection up to 500 A. Also, system constraints would limit module volume and power dissipation to approximately 500 cm^3 and 500 W, respectively. To achieve bi-directional protection, the well known dual common-source MOSFET structure shown in figure 2 was considered. However, the solution would have two times as many devices and twice the losses of a uni-directional blocking approach. The 1200 V, 20 A Cree SiC DMOSFET has a specific on-state resistance of 18 mΩ-cm^2 and a chip area of 0.18 cm^2.[1] The proposed 500-A / 500-W module voltage drop of one volt would require a drain-to-source voltage (V_{DS}) of 0.5 V per device, corresponding to a drain-to-source current (I_{DS}) of only five amps. Assuming reverse conduction could be achieved and a reverse voltage drop (V_{SD}) at or below V_{DS} for a given current, the proposed 500 A module would require at least 200 devices placed in two 100 device parallel strings with common-source connections. The study was refocused on other device candidates due to the difficulty in meeting the module requirements using SiC DMOSFETs.

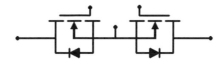

Fig. 2. Common-source structure in MOSFETs for bi-directional current blocking.

4. SiC JFET Approach

Presently, SiC DMOSFETs suffer from low MOS mobility and native oxide reliability issues.[2] Furthermore, several temperature dependant factors result in a decrease of the SiC MOSFET threshold voltage with temperature, which may lead to unwanted MOSFET turn on at temperatures over 200 °C.[3] SiC VJFETs are promising candidates for high-power and high-temperature switching as they only use *pn* junctions as a current control mechanism in the active device area, where the high electric fields occur. Therefore the high-temperature properties of SiC can be fully exploited in a gate voltage controlled switching device. SiC VJFETs have been successfully operated at 300°C with thermally induced parameter shifts in excellent agreement with theory.[4] This capability can benefit many military systems operating in high-temperature environments, and compact systems requiring the highest power densities.[5]

The normally-on SiC vertical JFET (VJFET) device developed by NGC, shown in figure 3, was investigated as a candidate for a bi-directional SSCB.   In purely fault protection applications, a normally-on function is preferred over a normally-off function. More importantly, a VJFET design can offer reduced losses and smaller chip sizes.   A NGC SiC VJFET having a total chip area of 0.1 cm^2, yields a specific on-state resistance of 5.4 mΩ-cm^2 for a V$_{GS}$ bias of 2.5 V, figure 4.[6]   NGC performed high-voltage measurements which showed V$_{DS}$ breakdown voltages above 1900 V with a V$_{GS}$ of -36 V, figure 5.   This performance enables a tradeoff between faster SSCB turn-off transitions and smaller snubber components, with the reduction of snubber size being more desirable.

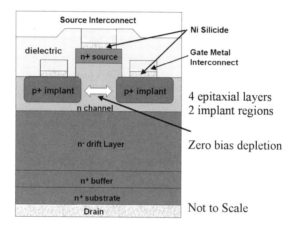

Fig. 3. Cross-sectional unit cell schematic of a normally-ON ion-implanted SiC VJFET (not to scale).

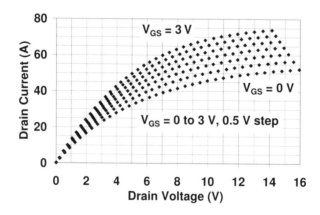

Fig. 4. Drain current versus drain voltage characteristics for a 0.1 cm^2 NGC VJFET.

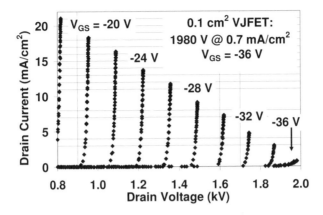

Fig. 5. Voltage blocking characteristics for a 0.1 cm² NGC VJFET.

Although the VJFET lacks a parasitic body diode between the drain and source terminals, it is not a symmetric device capable of bi-directional high-voltage blocking. Therefore, a series common-source connection of devices is still necessary for a bi-directional SSCB module. Bi-directional current conduction was evaluated for two 0.1 cm² VJFETs connected in a common-source configuration. Results are shown in figure 6. As expected, bi-directional VJFET current conduction is symmetric.

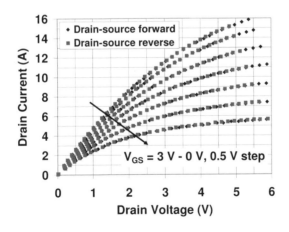

Fig. 6. Bi-directional current conduction of two common-source connected VJFETs is symmetric.

5. Small-scale Experimental Testing

Small-scale testing was done using 1200-V rated 0.03 cm² VJFET devices. A gate driver having adjustable output voltage levels was designed and built using a dual output isolated DC-DC converter feeding two cascaded adjustable linear regulators and an optically isolated driver stage. Individual 8 A chips were bonded in isolating packages.

Typical device conduction and voltage-blocking curves are shown in figures 7 and 8, respectively. To facilitate the initial characterization, the VJFETs were not encapsulated for high-voltage operation, which limited voltage blocking to below 300 V. A minimum gate bias of -20 V was chosen to be sufficiently above the recommended limit of -30 V to prevent source-to-gate breakdown. A positive gate bias of 2.0 V was selected to reduce on-state resistance and provide a 0.7 V margin below the 2.7 V forward conducting voltage of the gate-to-source diode. This $V_{GS} = 2$ V curve is outlined by the third data trend from the top of the forward conduction plot of figure 7.

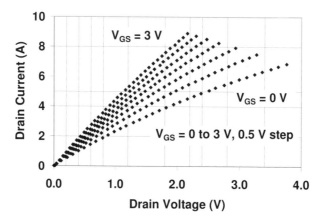

Fig. 7. Drain current versus drain voltage characteristics for a 0.03 cm² NGC VJFET.

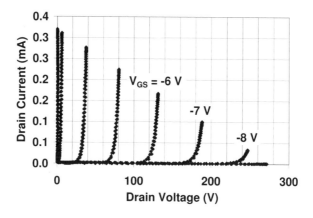

Fig. 8. Voltage blocking characteristics for a typical 0.03 cm² NGC VJFET without high-voltage passivation.

The parts were first evaluated in a simple test setup. Two 0.03 cm² VJFET devices were connected in a series common-source configuration with the bi-directional snubber connected between the two VJFET drain terminals. An isolated DC power supply was used as a source, and a 50 ohm load was connected in series with the VJFETs as shown in figure 9. The setup allowed the connections of the VJFET pair to be easily reversed for bi-directional testing. The gate terminal of the driver was connected to the gates of both VJFETs through individual five ohm resistors. The source terminal of the driver was connected to the common-source of both devices.

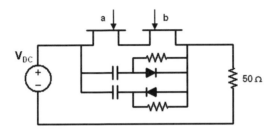

Fig. 9. Common-source VJFET test configuration.

For each test, the devices were initially held in the off-state at a V_{GS} of -20 V. Voltage on the DC power supply was then raised to the desired level. A gate pulse of 2 V was used to bias both devices to the on-state for five milliseconds, after which, the gate bias was returned to -20 V. Tests were run with currents from one to five amps through the devices in both directions. The one and two amp tests indicated symmetric bi-directional behavior. However, for tests run in either direction at or above 4 A, a reduction of the gate-to-source bias to a level below -20 V was seen following the device turn-off transition. During these tests, the 2.0 V rail of the driver was supplying continuous current (tens of mA) to the gate of the reverse conducting device during the on-state. The gate driver was not designed to provide continuous current to the gate of the device. As a result, the charge on the negative gate bias capacitor of the driver was increased by the level of continuous current being sourced.

It was determined that the continuous gate current was caused by the undesired forward conduction of the gate-to-drain diode. The larger currents conducted by the VJFETs resulted in larger voltage drops across the devices. For the reverse conducting device, a negative V_{DS} was established, which in conjunction with the V_{GS} of 2.0 V, forced V_{GD} to a positive ($V_{GD} = V_{GS} + V_{SD}$) value in excess of its built-in potential of approximately 2.7 V. This is schematically shown in figure 10, and is a basic manifestation of the Kirchhoff voltage law. The VJFET gate-to-drain diode forward conducts when biased above 2.7 V. The voltages between the drain and source terminals of the VJFETs, shown in figure 10, correspond to a current of approximately 4 A, which equals the load voltage of approximately 200 V divided by a load resistance of 50 ohms. In tests using a supply voltage of 250 V, an increase in the forward gate-to-drain current was observed. To avoid the undesirable turn-on of the reverse-conducting VJFET gate-to-drain diode, device biasing conditions must be modified for the series common-source VJFET configuration to function in a SSCB application.

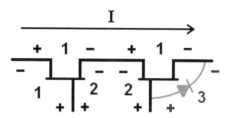

Fig. 10. Voltage distributions in the common-source VJFET test configuration can result in undesirable forward gate-to-drain diode conduction.

It is plausible that in series common-source VJFET blocking mode, the reverse conducting VJFET can reach a fully off-state before the forward conducting VJFET. This effect could damage the reverse conducting VJFET by exceeding its low reverse voltage breakdown limit before voltage is blocked by the gate-to-drain diode of the forward conducting VJFET. To address the issue, a second gate driver stage was added to the existing design to provide a delayed turn-on signal to the device in position to forward conduct (voltage blocking device). This same device was also provided a turn-off signal before the reverse conducting (non-blocking) device. The device pair was tested at current levels below four amps with delays ranging from two to ten microseconds. No adverse effects were seen. A graphical representation of the method of gate signal delay is shown in figure 11.

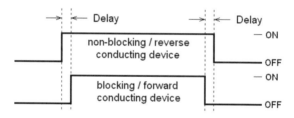

Fig. 11. Gate signal delays for bi-directional voltage blocking series common-source connected VJFETs.

As mentioned above, device operating conditions must be modified for the series common-source VJFET configuration to function in a SSCB application. Reducing or eliminating the positive gate bias of the VJFETs would increase the range of operating current but would also significantly increase device on-state resistance and power dissipation. Alternatively, to apply a reduced positive gate bias to only the reverse conducting device would increase gate driver complexity. Placing more common-source device strings in parallel to maintain a lower device reverse voltage drop, would require many more devices to be used. Even if enough devices were paralleled for a rated current, the resulting SSCB could reach a malfunction condition even at instantaneous peak currents at or approaching twice the module rating. This condition would be atypical of most semiconductor based switch modules.

6. VJFET Design for SSCB Application

NGC is presently designing a 0.1 cm^2 VJFET device with a more normally-on characteristic for common-source VJFET SSCB applications.[7] This VJFET can operate with no gate-to-drain conduction, in the common-source configuration, with minimal increases in on-state resistance and power dissipation. As the device is designed to be "very normally-on," a relatively small increase in VJFET specific on-state resistance is observed as the positive gate bias is reduced from 3.0 V to 0 V. Very normally-on VJFETs can also enable proper SSCB operation for short duration currents of over twice the module rating.

7. Proposed SiC JFET Device for SSCB

The design of a fully symmetric, normally-on, SiC device rated for 1200 V bi-directional blocking and bi-directional current conduction has been proposed. Such a device will operate bi-directionally in SSCB applications, eliminating the need for a common-source connection with an additional device. Reduced power dissipation and module size, as well as increased reliability could result from using such bi-directional devices. The device will require an unconventional gate driver circuit in SSCB operation. To evaluate its feasibility, a device design study would need to be conducted.

8. Conclusion

The increasing role of power electronic systems required by the military to support critical missions has prompted the need for more rapid and reliable fault protection modules. Limitations in conventional device technology and materials have shifted the focus to maturing majority carrier SiC device technologies. A normally-off module function is not always required, and a normally-ON function can be preferred in many applications. In such cases, the benefits of JFET devices can be exploited. These benefits include: low specific on-state resistance, lack of a parasitic drain-to-source body diode, high-temperature operation, and low gate charge. VJFET devices of 0.03 cm^2 were evaluated in a small-scale SSCB test setup. Operational limitations related to reverse conduction of the devices were identified and explained. New VJFET devices with more normally-on characteristics are presently being fabricated to improve performance in SSCB applications. Finally, a symmetric SiC JFET device has been proposed to offer additional performance benefits.

References

1. Ronald Green, Damian Urciuoli, Aderinto Ogunniyi, Gail Koebke, Lauren Everhart, Dimeji Ibitayo, Aivars Lelis, and Brett Hull, Evaluation of 4-H-SiC DMOSFETs for power converter applications, *2007 International Semiconductor Device Research Symposium.*
2. S. Krishnaswami, M. Das, B. Hull, S-H. Ryu, J. Scofield, A. Agarwal, and J. Palmour, Gate Oxide Reliability of 4H-SiC MOS Devices, *IEEE 43d Annual Intl. Reliability Physics Symposium*, 592-593 (2005).

3. D. Stephani, and P. Friedrichs, Silicon Carbide Junction Field Effect Transistors, *International Journal of High Speed Electronics and Systems*, **16**(3), 825-854 (2006).

4. V. Veliadis, H. Hearne, T. McNutt, M. Snook, P. Potyraj, and C. Scozzie, VJFET based all-SiC Normally-Off Cascode Switch for High Temperature Power Conditioning Applications, to appear *International Conference and Exhibition on High Temperature Electronics (HiTEC 2008)*.

5. D. P. Urciuoli, Power Dense Bi-directional DC-DC Converter Development, *2007 GOMACTech Conference*.

6. V. Veliadis, T. McNutt, M. McCoy, H. Hearne, G. DeSalvo, and C. Clarke, 1200-V, 50-A, Silicon Carbide Vertical Junction Field Effect Transistors for Power Switching Applications, *International Conference on Silicon Carbide and Related Materials*, October 2007.

7. V. Veliadis, M. McCoy, E. Stewart, T. McNutt, S. Van Campen, P. Potyraj, C. Scozzie, Exploring the Design Space of Rugged Seven Lithographic Level Silicon Carbide Vertical JFETs for the Development of 1200-V, 50-A Devices, *International Semiconductor Device Research Symposium 2007*.